MA 3 2410 01975 5191

RON PAUL'S REVOLUTION

———— ALSO BY BRIAN DOHERTY ————

This Is Burning Man

Radicals for Capitalism

Gun Control on Trial

RON PAUL'S RUTION

THE MAN AND THE MOVEMENT HE INSPIRED

BRIAN DOHERTY

BROADSIDE BOOKS
An Imprint of HarperCollins*Publishers*
www.broadsidebooks.net

HarperCollins books may be purchased for educational, business, or sales promotional use. For information, please write: Special Markets Department, HarperCollins Publishers, 10 East 53rd Street, New York, NY 10022.

Broadside Books™ and the Broadside logo are trademarks of HarperCollins Publishers.

FIRST EDITION

Designed by William Ruoto

Library of Congress Cataloging-in-Publication Data

Doherty, Brian
Ron Paul's revolution : the man and the movement he inspired / by Brian Doherty.—1st ed.
p. cm.
ISBN 978-0-06-211479-2 (hardback : alk. paper)—ISBN 978-0-06-211480-8 (trade paperback)—ISBN 978-0-06-211481-5 (ebook)
1. Paul, Ron, 1935- 2. Presidential candidates—United States—Biography. 3. Legislators—United States—Biography. 4. United States. Congress. House—Biography. 5. Libertarianism—United States. 6. United States—Politics and government—1989–2012 I. Title.
E901.1.P38D65 2012
973.931092—dc23
[B]

2011050691

12 13 14 15 16 OV/RRD 10 9 8 7 6 5 4 3 2 1

For Phil Blumel,

*All the College Libertarians at the
University of Florida, 1987–91,*

*And All Activists Helping People
Without Using Force*

CONTENTS

———— ★ ————

RON PAUL'S
R EVOL UTION

———— ★ ————

CHAPTER ONE

"RON PAUL: NOT GOING ANYWHERE, IDEOLOGICALLY PURE, AND TOUGH AS NAILS"

Rudy Giuliani, likely the next president of the United States, couldn't believe his ears. That scrawny old nut at the podium down the row had just said *what?* There was a *reason* America was attacked on 9/11? And it had to do with America's own behavior?

"That's really an extraordinary statement," Giuliani said, with some swagger. It was May 15, 2007, at a Republican presidential debate in South Carolina. Giuliani was in control. The former New York City mayor was front-runner of the Republican pack, and ahead of any likely Democratic opponent in the polls. But this barely polling former third-party candidate at the podium was attacking Giuliani's home turf—the 9/11 assault on America, and what it meant. "That's an extraordinary statement, as someone who lived through the attack of September 11, that we invited the attack because we were attacking Iraq. I don't think I've heard that before, and I've heard some pretty absurd explanations for September eleventh."

The crowd was on Giuliani's side, raucously, giving him cheers and whistles and resounding, rolling applause.

"And I would ask the congressman to withdraw that comment and tell us that he didn't really mean that," Giuliani continued.

If this obscure, unaccomplished, backbench legislator actually wanted to contend for the Republican Party's presidential nomination, Giuliani thought, he'd back down, pronto. Sputter some mealymouthed, face-saving scramble and hope this whole exchange was forgotten by the pundits and the voters.

That wasn't going to happen. The man, Ron Paul, from southeast Texas, then veteran of over nine terms in Congress looking too small for his suit, his ears almost laughably prominent, delivered his heresies neither hesitantly nor militantly, but with the authority of common sense. Paul knew what he meant, meant what he said, and given the chance, just explained himself further—or further dug his grave with the potential voters he was supposedly there to win over.

"I believe very sincerely that the CIA is correct when they teach and talk about blowback," Paul said. "When we went into Iran in 1953 and installed the Shah, yes, there was blowback. A reaction to that was the taking of our hostages and that persists. And if we ignore that, we ignore that at our own risk. If we think that we can do what we want around the world and not incite hatred, then we have a problem."

We have a problem? America doesn't *have* problems, pal—America *gives* problems! But this Ron Paul guy kept marching ahead into this dangerous, uncharted territory. "They don't come here to attack us because we're rich and we're free. They come and they attack us because we're over there. I mean, what would we think if we were—if other foreign countries were doing that to us?"

Whoa—a history lesson, recognizing consequences to our actions, an empathetic approach to what the rest of the world might think? What could any of that have to do with American foreign policy, or an attempt to win the Republican Party's presidential nomination?

This Ron Paul character hadn't been doing very well in his "quixotic" (that's what everyone said) run for the nomination so far in May 2007, and this surely was the end. Even many of his fervent fans, much as they enjoyed hearing him say it, were sure he'd just murdered his campaign.

One young man who thought he was past caring about electoral politics was idly watching the debate in Los Angeles. *This* made him sit up and take notice. "Ron Paul, without a fucking friend in the world, nothing but hostility aimed at him from all directions, stood his ground and did not back down. Just reiterated his points even stronger. I was blown away. I felt at that moment that the world changed forever, that there had been this massive shift in reality and what could happen. From that point forward I became involved." Jon Arden, who worked at Forest Lawn cemetery in Los Angeles, started donating money to the campaign and going to Ron Paul Meetups (local groups of Paul fans who'd gather both online and in the real world to promote Paul). Suddenly, making and hanging Ron Paul signs, talking about Ron Paul to anyone who'd listen and some who wouldn't, became Arden's passion. "All my time, money, anything I could spare, I devoted all of it to Ron."

Arden's was the most emphatic and impassioned version of that story I heard, but I encountered variations of it dozens of times from Paul fans: That spat with Giuliani, rather than derailing Ron Paul's progress, was the engine that propelled it to greater speed. This was the moment that turned Ron Paul from an easily ignorable distraction in the Republican race to, well, a more-difficult-to-ignore distraction. Still, most mainstream media and politicos continued to try to ignore him. But Paul's online poll results began to soar; the number of people watching and making videos promoting him on YouTube and joining his meetup groups zoomed. The second quarter of 2007, in which that exchange occurred, saw Paul raising only $2.4 million. But in the next quarter, after Giuliani supposedly dispatched him handily,

Paul pulled more than *twice* that, $5.3 million. And that wasn't the end of his momentum.

Ron Paul has been alive and kicking in American politics for a long time. He's served three separate stints in Congress as a Republican representative from Texas, beginning in 1976, and is still there now in 2012. He's even run for president before—with the Libertarian Party, in 1988. He came in third (but with fewer than half a million votes).

By any sober estimation, suggesting at a GOP debate, just six years after the airplane assault by radical Islam on American icons, that our foreign policy mistakes disturbed hornets' nests and we shouldn't be surprised we got stung should have meant Ron Paul would be alive and kicking no more.

Despite that moment's aura of legendary bravery to so many of his supporters, Paul remembers the spat with Giuliani lackadaisically. Paul has understood the world a certain way for a very long time, and not much surprises him. Having observed him since 1988, I'd say the only thing that's surprised him has been his own success in winning supporters as a presidential candidate. "My immediate reaction was, I couldn't care less," Paul says. "I'm here to tell what I think is the truth. I didn't think lightning was going to strike, that I *was* going to be president, but oh, this brought me down. It was just what I've been up against for thirty years. No different. It was just being verbalized in all the booing, but that didn't affect me. You know, I guess it's too bad they are booing me, but that's the way it is.

"People wanted to interview me right afterward, to ask me if I was going to drop out. What would I drop out for? They said this is the end for me. No one knew it was just the beginning. Kent [Snyder, his campaign manager] whispered to me, 'Guess what? You are winning the after-debate polls.'"

Giuliani, unwittingly, had helped launch the Next American Revolution.

That revolution has continued, past Paul's being trounced by John McCain in the race for the 2008 GOP presidential nod. Paul did outperform his sparring partner Rudy handily, though—Giuliani only managed to beat Paul's vote performance in three states.

Paul is a remarkably successful politician made of contradictions. Though a longtime Republican congressman, he's built his reputation on such wildly liberal stances as ending the drug war, halting wars in the Middle East, and scuttling the Patriot Act. Despite this, in 2010 and 2011 he's won the presidential straw poll at the Conservative Political Action Conference (CPAC), the seedbed of young right-wing activists.

He's got traditional conservative bona fides, too. He's for ending the income tax and killing the Internal Revenue Service, and for stopping illegal immigration; he also thinks abortion should be illegal. Despite this, right-wing politicians and thought leaders from Giuliani to Bill O'Reilly to the *Weekly Standard*'s William Kristol deride and despise him.

Paul's appeal is a curious mixture of populist and intellectual. He attacks the elite masters of money, banking, and high finance at the Federal Reserve and Wall Street. But his philosophy on politics and economics was forged through decades of self-driven study of abstruse libertarian economists such as Ludwig von Mises and the Nobel Prize–winning F. A. Hayek.

He's a staggeringly successful politician by some measures—the only congressman to win a seat as a nonincumbent three separate times. He continues to be reelected to the House election after election, almost always by a higher margin than the time before. He does this while violating most traditional rules of politics. He doesn't strive to bring home the bacon. His 14th District in Texas is highly agricultural, rife with rice and cattle farmers, but he always votes against federal agriculture subsidies. In a district with 675 miles of coastline, struck violently in 2008 by Hurricane Ike, he votes against flood aid and the Federal Emergency Management Agency—even calling for the latter's abolition on national TV. He vows to never vote for any bill for which he doesn't see clear constitutional justification. Yet by some people's standards of a "successful legislator" he's a bust—nearly every bill he introduces never even makes it out of committee.

For decades Ron Paul remained an underground hero to a national constituency of hard-core skeptics about government, the one successful politician steadfast even on the less popular aspects of the live-free-or-die libertarian philosophy. He'd talk about ending the drug war in front of high school students. In 1985, he spent his own money to fly and testify on behalf of the first draft registration defier to go to trial. Paul didn't blanch when confronted with the hot-blooded youngster's use of the phrase "Smash the state." He might not use that verb, *smash*, the sober obstetrician, air force veteran, and family man said. But from his experience with how the U.S. government disrespects its citizens' liberties, he understands the sentiment.

Paul's popularity has not waned since his presidential failure in 2008. It was *since* then that he began winning straw polls at CPAC. A national advocacy group pushing Paul's ideas, called Campaign for Liberty, arose from his campaign and raised $6.1 million in the off-election year of 2009—nearly three times what it raised in 2008. The organization Students for Ron Paul from that campaign evolved into Young Americans for Liberty, which now has 289 chapters and more than three thousand dues-paying members, and a network of twenty-six thousand activists to call on.

Giuliani was supposed to have killed him. John McCain was supposed to have killed him. But with Paul's predictions of trouble arising from America's overreach, foreign and domestic, seeming frighteningly prescient since the economic collapse of 2008—the continuing fall of the dollar, "peace candidate" Obama bogging us down further in Afghanistan, achieving an (incomplete) Iraq pullout only on George W. Bush's schedule, and starting a new war in Libya—Ron Paul is as alive as he's ever been.

Paul's supporters are alive and growing as well. His presidential campaigns have created the most lively, energetic, dedicated, and varied group of devotees for liberty that America has seen in living memory. They will cover the ground with homemade Ron Paul banners hung every place legal and illegal they can clamber; they will take to the air in blimps and balloons to promote their man; they will colonize and dominate every crevice of the Internet for him; they will ride their bikes across the country and turn from anarchist to Republican for him; they will run for office because he suggests they should; they will give more money, quicker, than any other political base in history. They are homeschooling Christians and couch-surfing punk rockers, college professors and famous actors, computer programmers and national TV hosts, drug-dealing anarchists and U.S. senators. They are the Ron Paul Revolution, and this book will explain who they are and how and why they are changing the shape of American politics.

I n late August, I was both writing this book and getting ready to go to the Burning Man festival, the subject of my first book (*This Is Burning Man*, 2004). On the surface, the two topics seem distinct—perhaps even hostile. Ron Paul is a Republican congressman and presidential candidate, a seventy-six-year-old Texas family man, fanatically free-market, opposed to ObamaCare and abortion and the welfare state.

In the popular stereotype (largely but not entirely accurate), Burning Man is a progressive-dream temporary community of hard partiers draped in absurd costumes and high on Ecstasy, making huge and unlikely art out of anything they can grab, celebrating a gleefully antinomian opposition to many aspects of capitalist modernity, and mostly eschewing cash-exchange commerce (though born of its excess).

I came across some news smashing those worlds together: the world of groovy giant art repurposing and détourning the detritus of capitalist consumer culture, and the world of the staid, sometimes stern politician who has been warning America for decades of the looming dangers of overextending itself through endless debt primed by the Federal Reserve's profligate practices.

In Kansas City, artist John Salvest had assembled 117 shipping containers and used them to build a huge wall in front of a local Federal Reserve bank. One side was emblazoned with the letters *USA*. The other side of the wall—the other side of the coin— read "IOU." Opposition to the Federal Reserve, once a Ron Paul monopoly in public life, had spread to a semi-mass movement of street protesters by 2009, thanks to Paul. By 2011, that distrust of the Fed was energizing public sculpture. Paul had done more than leave the imprint of his formerly peculiar ideas on politics; the culture itself was morphing from the force of Paul's powerful, if often occult, gravity field.

Ron Paul was everywhere. You wouldn't necessarily know it just from reading the papers and watching TV—and boy, do his fans know that. You can't spend more than a couple of minutes with them before they are complaining about media blackouts of their man. While researching this book, I had a dozen or more occasions to tell some civilian that I was writing about Ron Paul. In 90 percent of the cases, they had high awareness and an opinion, and 80 percent of the time that opinion was positive. Ron Paul sign wavers at the side of the road never fail to get an encouraging honk about every forty-five seconds, from Los Angeles to New Hampshire. By the rough and unscientific empiricism of moving through the world talking about Ron Paul, I found his national poll numbers, which continued to float through most of the summer and fall of 2011 in the 8–13 percent range, seemingly impossible and possibly even conspiratorial: in *my* world, it seemed everyone loved Ron Paul.

Paul was parodied on *Saturday Night Live* in September as a man so dedicated to the principle of governmental noninterference that he'd do nothing when faced with a house full of burning puppies, with bows on, their noses smushed up against the window, making sad puppy noises. He became a cause célèbre of *The Daily Show*'s Jon Stewart in the summer and fall of 2011. Stewart made a devastating bit out of the media's increasingly silly attempts to pretend Paul did not exist, noting that they treated this number-one-selling author, practically tied for first place in the election season's August kickoff straw poll in Ames, Iowa, like the "thirteenth floor of a hotel." After showing two absurd bits from Fox and CNN, in which commenters on the Ames results ignored Paul, the number two by less than 1 percent, tipping their hats instead to the importance of huge losers Rick Santorum and Jon Huntsman, Stewart flipped out with comic dudgeon.

"He is Tea Party Patient Zero!" Stewart shouted. "All that small government grassroots business, *he planted that grass*! These other folks are just Moral Majorities in a tricornered hat. Ron Paul's the real deal."

In a *Rolling Stone* cover story in September, though the interviewer never asked about Paul and wasn't interested in following up, Stewart kept bringing up the Texas libertarian. Stewart talked about the bit they do on *The Daily Show* where they show someone saying one thing, then something precisely opposite on another occasion. "You know a guy you'd have a hard time doing that to?" he says. "Ron Paul—because he's been consistent over the years. You may disagree with him, but at least you can respect that the guy has a belief system he's engaged in and will defend." Then later, "I don't understand how a guy with consistent grassroots support at the level he has is not a part of the conversation.

. . . Ron Paul has a constituency—like it or not, it's there. How can you just ignore it? It makes no sense."

Actor Vincent Vaughn was hanging out in Reno with Paul in September at a conference of libertarian activists sponsored by the Paulite organization Campaign for Liberty, each somehow making the other look cooler. Barry Manilow was telling the *Daily Caller* that "I agree with just about everything he says. What can I tell you?"

That same month, as old-school and mainstream a news source as *Time* magazine ran a feature dubbing Ron Paul, aptly, "The Prophet." *Time* noted that since Paul's last presidential run, "the world has changed in mostly grim ways that seem to affirm Paul's worldview. His vision of an eroding Constitution and a Washington–Wall Street cabal helped spark the Tea Party movement. Conservatives who once sneered at his foreign policy . . . have grown weary of war. His call for a more accountable and transparent Federal Reserve has morphed from quaint obsession to mainstream Republican talking point in Congress and on the campaign trail."

In October, *Saturday Night Live* revisited Ron Paul. Their attitude had changed, as the general GOP voting public's attitude was also starting to change; by late December Paul was polling ahead of all other comers in first-caucus state Iowa. In an absurd but vividly pro-Paul conclusion, Paul, exiled from an ongoing Republican candidate debate to a parking garage, is kidnapped by two guys who screech up in a white-paneled van. Paul dispatches his kidnappers with two off-camera gunshots from inside the van, brushes himself off, and returns to his lonely vigil. The fake debate host sums him up: "Ron Paul: Not going anywhere. Ideologically pure and tough as nails."

What he's done, why he's found such a fervent audience, and what Ron Paul's revolution means for the future of American politics are what this book is about. None of what has happened with and around him since that dustup with Giuliani had been conceivable by anyone—especially by Ron Paul. He'd been doing his thing too long, with too little to show for it, to believe his time would ever come. Being surprised about the unlikely rise of Representative Ron Paul is why I decided to write this book.

I should brag: I was into Ron Paul before Ron Paul was cool. I met him during his first run for president in 1988—with the Libertarian Party. It was January 1988, and a student group I worked with, the University of Florida College Libertarians, brought him to our campus in Gainesville to speak. I first wrote about him in 1999, in an article for the *American Spectator.* I watched his career from my perch as a reporter and editor at *Reason* magazine through the new century, and in 2005 I interviewed him in his D.C. congressional office for my book *Radicals for Capitalism: A Freewheeling History of the Modern American Libertarian Movement.*

I interviewed him frequently for *Reason,* and I was the first national reporter to interview him in January 2007 when he announced his presidential exploratory committee. That announcement was of little interest then to anyone outside the libertarian world that *Reason* covered.

Every step of the way as I covered his first campaign and then his second one, launched in 2011, I've underestimated how successful Paul would be. Not that I hadn't known him to be a rock star, in his way. But he was an *indie* rock star—the secret love of a savvy few, humble, of the people, too high-quality to be embraced by too large a mass. Brilliant, but esoteric. Or so I thought.

Then I saw the five hundred screaming college kids on Halloween weekend in 2007 at Iowa State University, cheering wildly during Paul's attacks on the Federal Reserve. Then he raised the largest amount of money in one day that any politician had ever raised, in a fan-organized "moneybomb" on Tea Party Day, December 16, 2007. Yes, this was long before the Tea Party movement shook American politics in 2009. Paul and his fans clearly helped shape, though they do not rule, that anti-bailout, small state mass movement. And then his fans took to the sky in a blimp advertising their man Ron Paul and his (their) revolution.

Then this longest of long shots, who everyone from George Stephanopoulos to your mom knew had no chance of getting anywhere with this run-for-president stuff, was the last man fighting it out with John McCain when all other Republican candidates had bailed, with too few votes and, most importantly, too little money once their "supporters" realized there was no way they could win. Those other candidates pretty much only had as much voter love as was commensurate with their realistic chances of winning; Ron Paul could keep raising money forever, and ended his campaign not in debt, but sitting on a pile of millions.

He did not win the Republican nomination. In fact, despite having at least twenty committed delegates choosing him for president, he wasn't even invited to speak at the party's 2008 convention. So this losing candidate decided to *throw his own convention.* The "Rally for the Republic" happened at the Target Center in Minneapolis while the GOP hunkered behind security fences in St. Paul, bedeviled by over ten thousand angry protesters. Paul meanwhile drew more than ten thousand adoring *fans* to his rally to (as the lefties say) not mourn, but organize. Ron Paul was no normal failed presidential candidate.

His success baffled even the man himself. By the normal standards of political success, Ron Paul does it all wrong. He has a unified and organic philosophy that includes at least one belief that's going to drive nearly every American crazy. He's as antiwar as any American politician can be. He's also against government health care. He thinks heroin should be legal. He also thinks abortion should be against the law. But to complicate things further, he doesn't think it should be against *federal* law. He doesn't believe crimes against persons like that should be a federal responsibility. He thinks the government should protect our borders and wants to end birthright citizenship and all government aid to undocumented immigrants. But he also thinks the border wall is a dumb idea and a waste of time. In 1983 he entered into the *Congressional Record* an essay by one of his intellectual heroes, Hans Sennholz. It reads in part: "it is futile to stem the human flood of immigrants with dikes of laws and regulations from the army of the police state. . . . In the cause of individual freedom, we must defend the rights of all people, including illegal aliens. . . . [I]f the political rights of an American citizenship entail the denial of the human right to work diligently for one's economic existence, and if we are forced to choose between the two, we must opt for the latter. The right to sustain one's life through personal effort and industry is a basic human right that precedes and exceeds all political rights."

Ron Paul wants to get rid of the income tax. He wants to get rid of the Patriot Act. He believes we should all be free to own and use guns as long as we aren't harming the innocent with them. He believes we should also be able to eat any food or drug or take any medical treatment we want, as long as no one but ourselves is harmed.

In conventional American political terms, this makes Ron Paul a heap of confusing paradoxes, difficult to sell. From his own libertarian constitutionalist perspective, this makes him about the only coherent, logical politician around, one whose stances can all be deduced from the premises he starts with: the premises that the United States of America started with.

You remember: that we are endowed by our creator with certain rights, including life, liberty, and the ability to pursue happiness. But remember, there's no guarantee, especially not from the government, that we are actually going to achieve that happiness. That's up to us, and to providence. We have the right to own and use property as long as we aren't directly physically harming someone else. We have the right to speak and act in any manner we please, as long as we aren't directly and physically harming someone else. Paul himself is a man of great cultural conservatism—never even seen pot consumed, barely drinks, a staunchly religious family man who nonetheless refuses to crow about it for political advantage. Yet drug-dealing heathen tell me with grave seriousness that they would take a bullet for Ron Paul, because he is that important—to them and to the nation.

That's Ron Paul. From his own perspective, a classic American patriot. To many others he's at best a perplexing anachronism, at worst a frightening menace. Right-wing pundit and former William Buckley protégé Richard Brookhiser calls Paul's fans "wicked idiots"; popular right-wing website RedState tried to ban them; Mona Charen, former speechwriter for Jack Kemp and Nancy Reagan, wrote that he "might make a dandy new leader for the Branch Davidians."

Paul's rigorous hewing to a vision of government that almost every part of America's learned political, academic, and media elites considers silly was only the start of his problems with the American electorate. While most politicians treat Americans as fragile brats who need to be constantly assured how great they are and how great things are (except for some small problems that the *other guy* caused and are easy enough to fix by *me*)—that we are the richest, freest, most righteous nation around—Ron Paul is here to tell us that our fiscal mismanagement and crazy debt have created a mere illusion of prosperity, that an economic crisis even worse than 2008 is nearly unavoidable if we don't change our ways and live within our means, pronto.

Paul will tell us how the government no longer respects or protects our rights but instead violates them—sending armed agents to bust in on people selling raw milk to willing customers, physically groping and probing us before we can travel on planes, murdering us on presidential ukase. He will lecture us that the way we treat foreigners overseas is not righteous, is not proper, is not in defense of our own liberties, is actually the criminal behavior of a decadent empire, and we shouldn't be surprised when all that creates resentful enemies. Most shocking, he will tell us we should not be so surprised when those resentful enemies try to strike back, in their small, lacking-a-military-bigger-than-all-the-rest-of-the-world-combined way.

It's not comforting, Ron Paul's world. It doesn't tell us we are good and right and strong and that everything will be okay. It can be scary, though often the apocalyptic undertones are something you have to figure out for yourself—Paul doesn't always shove your face in it. But he's got a coherent, educated view of how economies work that convinces him that our economy has even worse troubles ahead. This is not some ad hoc idea that a team of advisors he just hired taught him a month ago. It's a viewpoint he's been marinating himself in, even before he was a politician, for forty years now.

He knows you've probably never heard of it, but he gives it a shout-out every chance he gets. Paul's preferred brand of economics is called the Austrian School. The names Paul is most likely to drop are Ludwig von Mises and F. A. Hayek. Studiousness about things no one else knows or cares about isn't a typical path to political success, either.

His Austrian understanding of economics told Ron Paul—long before you or anyone else in power in Washington knew it—that the seeming prosperity of the housing boom of the last decade was phony, ginned up through untoward credit creation by the Federal Reserve via Alan Greenspan's reckless dedication to unnaturally low interest rates. When he's got sixty seconds he'll try to explain this whole boom-bust cycle, and why gold money instead of paper money is the way to solve the problem. It doesn't really work, but he's trying. Again, not the sort of thing a politician does when trying to win voters' love.

All of the above is why I, and the world, have been surprised by Paul's ability to win dozens of straw polls, and poll roughly equal to or beating Obama in most 2011 one-on-one matches with the president. Paul is as surprised as the rest of us. He knows his reputation as a kooky outsider, and he'll play with it in a cute but sly way. Things are getting so crazy, he'll tell friendly audiences and elicit a knowing laugh, that they are telling him he's becoming mainstream!

He's still not. But the Ron Paul Revolution did not die after 2008, did not falter in the age of Obama. This book will tell the story of Ron Paul, his ideas, his people, and why despite failing to win the Republican Party's nomination again, he is likely the only politician on the scene today who will shape the future of America in a significant way. The energy of his devotees, and the seriousness of the fiscal, monetary, and foreign policy crises facing us, ensures that Paul's ideas will march on even when Paul, who is leaving his congressional seat in 2013, no longer does.

This book is not out to sell any one Big Idea about Ron Paul. Making such generalizations about the whys of national political or cultural trends requires eliding or ignoring too much. In this very evading of Big Ideas is a Paulian idea: ideological and activist individualism. Ultimately, ideas catch fire because many thousands of individuals choose to accept them. The libertarian conspiracy of which Paul is the latest and most successful example, dating back to movement founding father Leonard Read, who launched the first modern American libertarian think tank, the Foundation for Economic Education (FEE), is dedicated to that slow, deliberate changing of individual mind by individual mind.

Unlike many other American political movements, Paulism represents the interests of no particular region, no particular economic sector, no particular specific oppressed class or specific elite. But though no one Big Idea explains Why Ron Paul? and Why Now?, any Big Idea about American politics today and tomorrow that ignores Ron Paul and his devotees is missing so much of importance that it isn't worth much. Paul's position in American politics seems curious and confusing to many. But he's actually in a grand American tradition of radical Jeffersonians and currency obsessives who capture the devotion of a mass minority in American history, and send the nation slaloming off in new directions.

———— ★ ————

CHAPTER TWO

RON PAUL IN THE
AMERICAN GRAIN

Paul's rise in prominence as a public figure in America hasn't dispelled the fog of ideological confusion that surrounds him. Paul isn't confused himself. He knows who he is, what he believes and why. He fully grasps the internal coherence of his combination of classic American constitutionalism that keeps government powers listed and limited, in the spirit of the Founders; belief in a unified personal and economic liberty based in natural rights; Austrian economics that teaches him that free markets and hard money are the only guarantee of a prosperous economy that reflects the desires of the masses; and a practical concern with the abuses of powerful elites out to harm the American people to their advantage, whether those elites are government or corporate.

Paul's beliefs make partisans of both left and right see Paul as their sworn enemy, a dangerous reactionary kook the likes of which they can't seem to believe has arisen and found traction in the modern American context. But while his ideas and style may have been out of fashion for a long time, they have a long, if strange, pedigree in American politics.

The particular combination of beliefs that animates Paul and his fans has not been prominent on the American scene since the Locofocos. You don't remember them, but they were the New York–based, radically laissez-faire wing of the Jacksonian Democrat coalition during the President Martin Van Buren era of the 1830s–'40s. Like Paul, for them, the separation of government from banking was their primary goal (as well as the elimination of non-hard money). But aspects of Paul and Paulism appear and reappear across our nation's history, like ghosts haunting the battlefield where the American dream has been slaughtered in slow motion since shortly after it was born.

Paul's valorization of liberty, his hatred of central banking, his anti-imperialism, and his populist-insurgent style opposed to academic and political elites are all enduring parts of the American tradition. Paul's style is that of the amateur intellectual selling a painstakingly coherent alternative explanation of the way the world works and should work. It's a style derided by doubters as "crackpot." Still, Paul's spirit has been embraced over the years by passionate minorities in our land of majority rule. Sometimes such movements change America, ineluctably and powerfully. Sometimes their spirit, arising out of time either too early or too late, fizzles. But such movements map the frontiers where the ideas that will shape America tomorrow fight for respect today.

America likes to claim a unity that isn't quite real: that it is this one thing, means that one thing. But Americans have many myths, many traditions, many equally essential spirits. Our experiences and our ideologies are as huge and variegated as the races of man that have flowed into and filled our capacious spaces, hewed and shaped our forests, dammed and sailed our rivers. Our Civil War reifies an essential truth: We are within our unity always and also at least two in conflict. As much as Ron Paul's political opponents want to pretend that the tradition he represents and continues is strange and laughable, some curious alien infection, more funny than serious, it isn't so.

Paul's methods of explaining himself, for all their obeisance to the Constitution, don't in practice rely that much on appeals to Founding Fathers or ancient history. He's more likely to appeal to common sense, justice, and the GOP's own twentieth-century history. Paul is enough of a purist about liberty to recognize that it has more of a future than a past, that the movement he leads is less about returning to some perfect past than about perfecting (to the degree it can be perfected—he's no utopian) tomorrow. During 2011 campaign speeches, he talked of how we need to extend and perfect the traditions of liberty, not pretend there was a past golden age to restore. Yet, he will also sometimes say America's practice of liberty went awry one hundred years ago, in the Progressive era, including World War I, the income tax, and the beginnings of the modern corporatist state. (Given the social progress, and the improved legal status of women and blacks, since then, many modern libertarians face-palm themselves at even a hint that the nineteenth century was some wonderland of liberty worth remembering fondly or fighting to return to.)

Paul is as all-American as he thinks he is. But so was his antipode Franklin Delano Roosevelt. However strange and displaced he seems in current American politics, Paul is in the American grain, even in how he is statesman to his fans, crackpot to his foes. Civil war and an electorate split furiously down the middle are right in the American main.

Even as the United States of America was being born, two opposing forces fought over the purpose and shape of our political union, the Federalists and Anti-Federalists. The Federalists were scared of the chaos of a multitude of independent states and an untaxed rabble. They won, by imposing on the several states a unitary Constitution, whose radically unitary nature was proven and forged a few decades later, in the Civil War. In 1789, the supporters of the Constitution were the Federalists; those who didn't trust it, the Anti-Federalists. Ironically, those who today promote a government restricted to its constitutional limits—that is, those on the side of the Constitution, like the original Federalists—would almost certainly have been Anti-Federalists in that time. Modern constitutionalists such as Ron Paul want to keep the government restricted to powers set out for it in the Constitution for the same reason the Anti-Federalists were leery of that document to begin with: they mistrusted men with power, and wanted them bound. They sought not national greatness, like so many modern conservative intellectuals, but local (and individual) liberty.

The Federalists argued that the Constitution would create a limited government, respectful of rights, one that honorable men need not fear; the Anti-Federalists didn't believe it. The Anti-Federalists turned out to be right. States and localities lost their unique powers; the federal government's control over them and the people metastasized wildly. The best that the Anti-Federalists' ideological descendants today can hope for is that the U.S. government can be made to function the way the Federalists promised in the beginning. The Anti-Federalists are now the constitutionalists, and the Federalists are those who think that the general welfare and commerce clauses cover not just a multitude of governmental sins, but all of them.

Ron Paul sees many enemies in modern government, but the one he has grappled with most directly and persistently is the Federal Reserve, the modern equivalent of the Second Bank of the United States, famously slain by Andrew Jackson. Jackson's reasons for killing the bank were Paulian. As Jackson's most famous twentieth-century analyst, Richard Hofstadter, put it, the Jacksonian movement was "essentially a movement of laissez-faire, an attempt to divorce government and business." Jackson hated the bank because it was secretive and corrupting and inflationary, all the same reasons Paul hates the Fed now.

Jackson biographer Robert Remini notes that Jackson was incensed that the bank refused a thorough investigation of its affairs, just as Paul has tried (and failed) to audit the Fed for decades. Jackson was a man of passions more than coherent philosophy, which made him a more successful politician in a mass democracy than Paul could likely be. But the spirit of Jackson's veto message of the bank, as Remini sees it, is the sort of thing only Paul (or possibly Ralph Nader) among modern national political figures could write today: it, as Remini explained, "powerfully restated the philosophy of the minimized state. Centralized government endangered liberty . . . and therefore must never intrude in the normal operations of society. When it does interfere and assume unwarranted power, such as creating a bank, it produces 'artificial' distinctions between classes and generates inequality and injustice." Paul himself, in his 1982 book *The Case for Gold*, claimed the Jacksonians as ancestors. He wrote that despite being "flagrantly misunderstood" by historians, "the Jacksonians were libertarians, plain and simple. Their program and ideology were libertarian; they strongly favored free enterprise and free markets, but they just as strongly opposed special subsidies and monopoly privileges conveyed by government to business or any other group."

Mistrust of central banks and the paper money that the Federal Reserve creates is an American tradition going back to the collapse of the Revolutionary War–era paper continental, as in "worthless as a continental." Lincoln's first attempt to issue a greenback in the Civil War was first declared unconstitutional (which it was—as Paul frequently points out, Congress' constitutional power to "coin" money does not seem to allow room for paper) before a later court gave in to political expediency, as later courts always do. The fight for stable money and against the inherent power imbalances of huge national banks that Paul continues was begun by Jackson, though few others have kept that fight going since Jackson's time.

By striving to make the currency question—what is money,

and who should control it?—central to American politics again, Paul is reviving and echoing a debate that was far more prominent in nineteenth-century America than most remember. From the Jackson-era struggle over the Second Bank of the United States through the Greenback movement and the Populist era, the national and sectional debate over what money should be (roughly, metal or paper?) and what forces should guide it (roughly, markets or politics?) generated heated debate and much heavy thought, pamphleteering, and activism. The money debate shaped presidential contests and inspired many third parties, especially after the Civil War, that were more successful than third parties ever are these days. And, as Ron Paul could have warned them had he been around, America's money problems reached a height with the excessive debt and paper money creation necessitated by war—in the nineteenth-century case, the Civil War.

As historian Gretchen Ritter wrote in her great 1997 survey of these nineteenth-century debates, *Goldbugs and Greenbacks* (Cambridge University Press), the politics of currency were as serious and dominant an issue in the second half of the nineteenth century as slavery had been in the first half. And even though Ron Paul ultimately disagreed with the Greenbackers and Populists of the late nineteenth century about whether it was better that money be metal shaped by markets or paper guided by government, Paul today in his adumbrations against the Federal Reserve and its secret bailouts of its banking and finance buddies echoes Populist-era complaints that big private banking interests had too much power to aid their own business, class, or sectional interests to be left unrestricted. (Paul is not a devotee of theories that the Federal Reserve is bad *because* it is in some sense private, which it isn't in any effective sense. Paul thinks the Fed is bad because of its governmental role as the engine of inflation and crony giveaways.)

The major American political figure Paul most closely and

curiously echoes arose from the late nineteenth-century ferment against privilege and empire, and was far more successful than Paul has been, winning the Democratic Party nomination three times—yet failing to win the presidency each time, too.

He is William Jennings Bryan. Casual acquaintance with the one big idea associated with the two would make them seem insuperable rivals. Ron Paul wants a gold-based, noninflationary monetary system. William Jennings Bryan made his name as the defender of cheap and abundant money for farmers and debtors, famously declaring that it is a sin to crucify mankind on a cross of gold. The two men seem unalterably opposed on their central issue. But they are brothers across American policy history in knowing how important that issue is. Both understood that money and banking and their operations were central to the health of the economy and the people in ways that most politicians don't understand—even if Paul believes that keeping the power to make more money willy-nilly out of the hands of government, which he insists is good for everyone, was more important than direct action to help debtors.

But Paul could still have seen Bryan as a brother—a brother confused about economics, but no lover of central control and management for its own sake. The two politicians' character, style, populist appeal, and most specifically their anti-imperialism mark them as kin. Their similarities in style are summed up well by Bryan biographer Louis Koenig, who wrote that Bryan "was an ideologist devoted to a body of serious political beliefs that were germane to society's central problems, and he was willing to place them above victory." Another Bryan biographer, David D. Anderson, makes Bryan sound even more like Paul: "He was determined to persuade the nation that the principles upon which the country had built its independence, those phrased by Jefferson more than a century before, were neither mere rhetoric nor an abstract ideal but a valid, workable way of political, social, and economic life." When Bryan had a chance in the Senate to fight imperialist expansion with procedural means that allowed a minority to block majority will, he didn't want to: he wanted to win over the majority by changing their minds.

In the fight against imperialism, Bryan was even bolder in rhetoric than Paul. Paul only reminds us that our actions motivate our enemies to attack. Bryan, when he was accused of giving our enemies encouragement while opposing our occupation of the Philippines, said of the Philippine rebels fighting the United States that, as Bryan biographer Robert W. Cherny summed up Bryan's message, they "needed no encouragement from any living Americans for they could find ample inspiration from such Americans as Patrick Henry and Thomas Jefferson."

Bryan arose in a late nineteenth-century Democratic Party dominated by Grover Cleveland (Ron Paul's favorite president for his explicit recognition that he was not meant to be all-powerful, which Cleveland would explicitly say when vetoing do-gooder bills he saw no constitutional basis for). Cherny describes the party of that time as an "awkward coalition [that] sustained itself primarily because the members agreed that they did not want government to do much beyond deliver the mail." Bryan and Paul were both figureheads for national organizations promoting their ideas (in Bryan's case, Free Silver Clubs; in Paul's, the Campaign for Liberty and Young Americans for Liberty). Both followed up a failed presidential run with a huge bestseller that helped cement their fans' attention and affection for next time (Bryan's *The First Battle*, Paul's *The Revolution*). Both were overwhelmingly funded in their races by small individual contributions, not corporations or the superwealthy. Both were personally religious and abstemious. Both were loved and supported by political movements and third parties outside the major parties they ostensibly represented (Bryan's Populists, Paul's Libertarians).

In the run-up to World War I, Bryan, even as secretary of state, sounded like Paul in gauging his own country's actions against standards of neutral justice and empathy. After the submarining of a British ship, Bryan declared that America should not consider the drowning of a few men to be worse than the starvation of a nation by blockade. Both Bryan and Paul were moral absolutists more than they were detail men, and both considered the success of their principles more important than electoral success.

Bryan's more populist political stances—the ones he did not share with Paul, such as state management and regulation of businesses he saw as public trusts—succeeded without him succeeding per se. But it would be wrong for the historian to conclude that electorally feckless, uncompromising, widely derided figures such as Bryan (and Paul) are unimportant. Certain ideas survive the unsuccessful politicians who keep them alive. Contemplate, for example, Bryan's second presidential nomination acceptance speech, in 1900. It could come almost verbatim from Paul today:

> Behold a republic, resting securely upon the foundation stones quarried by revolutionary patriots from the fountain of eternal truth—a republic applying in practice . . . the self-evident propositions that all men are created equal; that they are endowed by their creator with inalienable rights, that governments are instituted among men to secure these rights, that governments derive their just powers from the consent of the governed. Behold a republic in which civil and religious liberty stimulate all to earnest endeavor and in which the law restrains every hand uplifted for a neighbor's injury—a republic in which every citizen is a sovereign, but in which no one cares or dares to wear a crown. Behold a republic standing erect while empires all around are bowed beneath the weight of their own ornaments—a republic whose flag is loved while other flags are only feared.

For Paul's own part, in his *Case for Gold* he characterized Bryan's rise as representing the death of the Democratic Party as the last quasi-laissez-faire political grouping with national power, replacing that with a Christian Pietist obsession with using government as a tool to "stamp out personal and political sin" and using government as a tool to impose a particular vision of virtue. That is not what Paul thinks government ought to be.

The age of Woodrow Wilson destroyed a lot of what Paul loved about the Old Republic—we were now overregulated, overtaxed, and over there. Paul is the last man standing of what those few political historians who note them at all call the Old Right, who kept alive a pre-Wilsonian vision of the nature and purpose of government even into the New Deal years. The Old Right was the movement of the *Saturday Evening Post* and the *Chicago Tribune*, of the America First movement and the Liberty League, of a mass of nonplutocrats who mistrusted the centralized control and foreign interventionism of the New Deal. H. L. Mencken was its most popular representative, though he didn't think of himself as representing anything per se other than a love of liberty and a hatred of Puritanism, busybodyism, and cant. The Old Right hated the New Deal for its regulation, for its income transfers, for its regimentation, and for forcing the Supreme Court to agree with an interpretation of government powers that was grossly unconstitutional, and they hated President Roosevelt for, in their minds, helping maneuver us into a European and Pacific war that shouldn't have been ours to fight.

Paul remembers fondly a congressional forebear almost entirely forgotten: Representative Howard Buffett (R-Neb.), the last thoroughgoing Old Right man in Congress from the original school, who entered Congress as an opponent of the New Deal and war during Roosevelt's last term. Buffett was both taught by and promoted the works of FEE's Leonard Read, one of Paul's libertarian guiding lights. (Buffett is best remembered these days for being the father of superinvestor Warren Buffett.) Buffett served a total of four terms from Nebraska; he also served as Midwest campaign manager in Robert Taft's ill-fated attempt to win the GOP nod in 1952 over Eisenhower. This was the famous last gasp of Paul-style small government and humble Republican foreign policy, since the Cold War soon buried it.

Buffett was personal friends with one of Paul's libertarian mentors, Murray Rothbard; the anarchist Rothbard considered Buffett and Paul the only truly good congressmen of his lifetime. Buffett, like Paul, was prescient about the costs and complications of Middle East interventions. He inveighed in 1944 against building an oil pipeline in Saudi Arabia with American money: "This venture would end," Buffett said, "the influence exercised by the United States as a government not participating in the exploitation of small lands and countries. . . . It may be that the American people would rather forego the use of a questionable amount of gasoline at some time in the remote future than follow a foreign policy practically guaranteed to send many of their sons, if not their daughters, to die in faraway places in defense of the trade of Standard Oil or the international dreams of our one-world planners."

It may be, but it may not be; we haven't really been given much of a choice. That, in 1944! Buffett predicted the Cold War's dire effects for American liberty and treasure as it was born. And he knew, again in a precise preview of Paul in the age of Islamist terror, that "patriots who try to bring about economy would be branded as Stalin lovers. The misery of the people, from continued militarism and inflation, would soon become unbearable." Buffett fought, but failed to stop, the draft in the 1950s.

Paul first came into Congress one year after another congressman whom he closely resembled was leaving it. This was H. R. Gross of Iowa, known as "Mr. No" before Paul became Congress's "Dr. No." Gross was against almost all spending (except agriculture appropriations for his state—he lacked Paul's all-round integrity on that) and was known for leading off his talks about any given proposal with "How much will this boondoggle cost?"

Like Paul, Gross's real effect, as one of 435 with a peculiar set of intentions that marked him, in Gerald Ford's estimation, as essentially his own third party, was small; you can't be an "effective legislator" that way. But Paul supporters in Iowa say Gross is still fondly remembered in his home state, in a way more generic and well-behaved congressmen of the past are not. As a campaigner, Ron Paul does almost no localized glad-handing, but he would go out of his way to summon the spirit of H. R. Gross when campaigning in Iowa in 2011.

The most direct analog of Paul's role in the Republican Party today is Barry Goldwater in 1960—a populist insurgent with a huge young fan base fascinated with their man's integrity in the cause of limited constitutional government, a young base attempting to infiltrate a party apparatus that doesn't know what to do with them and isn't sure it wants them, author of a national bestseller, trying to make his way in a Republican Party dominated by old, "respectable" thinking that sees the insurgent as crazily radical.

The analogies between the two, especially the ideological ones, are nearly too obvious to belabor, and Paul people think of them all the time today when told that someone as radical as Paul could never shape the future of the Republican Party. Foreign policy, as ever with Paul, is where the analogy fails: Goldwater was famously bellicose toward the United States' perceived enemy of the time, the Soviets, and Paul's lack of bellicosity may prevent him from achieving what Goldwater achieved. (Paul also lacks the experienced and talented state and national coalition-building machinery that Goldwater had behind him from his Arizona beginnings; Paul is still pretty much pure insurgent.)

From the other side of the party line another Paul-like insurgent arose in 1968: Eugene McCarthy, the quiet, intellectual, antiwar senator, cranky about the income tax, who inspired radicals to go "clean for Gene" in his fight against an unjust war that his party loved. McCarthy talked a tough moral line about America's choices and was vilified as being on the side of America's enemies. You might see a pattern: the further from Andrew Jackson you get with Paul's forebears, the less concrete political success you see. However, in his losing, McCarthy did help encourage Lyndon Johnson to give up on the presidency—a fair accomplishment.

Paul and the strategists around him seem to have a more recent, and more accomplished, pair of populist-leaning outsiders in mind when they imagine his future, if it is not in the White House: the resonant outsiders of 1988, Pat Robertson of the right and Jesse Jackson of the left.

Both were more successful on the primary level than Paul in 2008, and both men created organizations and instructed their followers to inhabit and change their parties, launching a crusade meant to outlive the particular ambitions of the candidates who initially led them. Robertson in the GOP started from a better position; his enemies, in the case of 1988, George Bush, understood his importance and tried very deliberately to co-opt his audience, even hiring some of Robertson's people on to Bush's own campaign. Allen D. Hertzke, a historian of both Robertson and Jackson, indicates the most encouraging way forward for Paulites who imagine their influence on the GOP can be as strong in the next ten years as Robertson's was from 1988 to 1998: "Far more than other candidates, these two ministers led social movements that continued to wield influence long after the 1988 election was over. Jackson and Robertson formed organizations dedicated to registering and mobilizing voters, educating supporters, and holding candidates of their respective parties accountable on issues of concern. Both focused their efforts on increasing leverage in their respective parties."

Paul's movement lacks the vivid sense of threatened cultural or group identity—whether Christian evangelical or black—that helped forge and empower the Christian and Rainbow coalitions. All three candidates' bases can rightly claim that they are struggling against an elite party apparatus that disrespects their needs, a resentment that can be quite politically motivating, though it doesn't always lead to victory. Still, in Robertson's case at least, his people effectively played the game of fighting their way into and against a GOP apparatus that was first unimpressed and annoyed by them, and so achieved party influence beyond their national numbers. Hertzke's account in his 1993 book *Echoes of Discontent: Jesse Jackson, Pat Robertson, and the Resurgence of Populism* of the mistrust and difficulties and often contempt that the incoming Robertson forces received from the existing party regulars echoes almost precisely stories one hears today from Paul activists entering the Republican Party.

Paul's liberty coalition lacks not only the predetermined group identity of the Christian or Rainbow ones; he doesn't have an inch of the pietism about personal behavior or choices of both Robertson and Jackson; Jackson the pastor was more of a moral traditionalist in his time than most remember. That lack might hurt Paul as a political figure. While Americans are attracted to liberty, they can also get very politically attracted to people telling them how badly some other group is behaving and how something must be done about it.

A recent Republican Party rebel whose ideas and interests most closely intersect with Paul's is Pat Buchanan, the anti-empire, antiestablishment Republican of the 1990s. Though they shared an aversion to post–Cold War American military adventurism, and a populist disdain for Washington, D.C., elites managing the country, Buchanan was no libertarian, and Paul is no culture warrior.

Both object to international trade agreements—Buchanan because he is a protectionist, Paul because he is a purist free trader. But for those concerned most with matters of war, the two seemed almost interchangeable in the 1990s. In 1992 a group of libertarians aiming for right-populist appeal tried to get Paul to run for president again but hitched their wagon to Buchanan when he announced his intention to seek the GOP nomination. While Buchanan did better than Paul in Republican primaries—in his second run in 1996 he won four states—he did not conjure a new wave of young grassroots enthusiasts willing to continue the fight; today barely a trace of Buchananism lingers in the political air.

Paul's devotees, sensing the impossibility of him winning the Republican nomination, urged him to do a third-party run, both in 2008 and 2012. In what might stand as a lesson, Buchanan's Reform Party run in 2000 ended his political career in an embarrassing way, with a fourth-place, 0.4 percent showing.

Despite the handful of policy similarities, Buchanan played consistently to old resentments about class and culture, unlike the Paul of 2007 to now. Forging a youthful army was not something Buchanan would ever be able to do, or be comfortable with even if it happened. Paul's lack of pugnaciousness, lack of connection with villains such as Buchanan's old boss Nixon, and his having abandoned the cultural pugnaciousness and nastiness associated with the populist right who loved both men in the 1990s, has ensured that Paul and his movement will have a future in both the Republican Party and America that Buchanan does not.

———— ★ ————

CHAPTER THREE

DR. NO GOES
TO WASHINGTON

Ronald Ernest Paul was born in 1935, the son of dairy entrepreneur Howard Paul and his wife, Peggy. He was the third of five sons, born and raised in Green Tree, Pennsylvania, on the outskirts of Pittsburgh. Paul's grandfather Casper was a teen when he arrived in America with his father, Paul's great-grandfather, who died within a year. Casper settled in Green Tree and started a farming business, and later a dairy; Casper's son Howard inherited the dairy business when his elder brother launched a farm supply chain.

Ron Paul grew up in a serious family, all about work and church. The five Paul boys and their uncles all worked the family dairy. The boys would inspect the bottles their uncles cleaned. Two of Paul's brothers became ministers, and he considered doing the same. (The other two became a math professor and an accountant.)

Paul navigated American small-town childhood with aplomb: student council president and star athlete, excelling in track until a football-induced knee injury slowed him down. (The injury persisted, and Paul's knees are now artificial.) Paul switched his athletic energy to swimming. His high school years overlapped with the Korean War—the first major undeclared war of American modernity. One of Paul's high school coaches went into that war and never came out. Having his childhood and adolescence unfold during World War II and Korea, and seeing neighborhood and family friends leave and not come back, imbued Paul with a disdain and distrust for war, and as he once said, a preference for "healing to maiming, life to death." He remembers a neighbor later who was finishing law school: "his dad was a lawyer, he went over, got drafted, killed in Vietnam. I remember going to his funeral, and it made no sense," Paul says. "And it made less sense as the years went on. The wars weren't necessary, and it helped mold in me a strong abhorrence of the uselessness of wars. And the more I looked at history, the stronger I felt about the whole issue."

Paul went to a small, church-affiliated college not far from home, Gettysburg College. He continued to be a model young American man, president of his fraternity's pledge class, Lambda Chi Alpha. He worked as many jobs as he could, including managing a campus coffee shop. He drove for his family dairy during the summer. (Dairies still made home deliveries in Ron Paul's America.)

He married his high school sweetheart, Carol Wells, during his senior year of college in 1957. Carol invited *him* on their first date. Jackie Gloor, who runs Paul's congressional office in Victoria, Texas, told me of Ron Paul's reputation as the man whom the Speaker of the House can't tell what to do, whom the president of the United States can't tell what to do. Who can tell Ron Paul what to do? Carol Paul.

Paul is practically from Pittsburgh, but neither he nor Pittsburgh makes much of this. Bill Steigerwald, a lifelong journalist in the Pittsburgh area, accompanied Paul on one of those "candidate visits the old hometown" TV segments in 2007. "Nobody had a clue who he was or would have recognized him. He was just this strange guy with a camera crew, no semblance of politics. He was nice as could be, just this old schmoe around the neighborhood twelve minutes from downtown Pittsburgh.

"Pittsburgh usually makes a big deal of its natives," Steigerwald says. "They always like to let you know that Michael Keaton is from Pittsburgh, Gene Kelly. But Ron Paul has mostly been forgotten by Pittsburgh."

Paul earned a bachelor's degree in biology from Gettysburg in 1957 and went to medical school at Duke University in North Carolina. It would be a mistake to make too much of his childhood and upbringing in helping to forge his unique politics. Like so many libertarians, Paul came to his politics more from philosophy than from experience, from the contemplation of the right (or wrong, depending on your perspective) books, not in direct osmosis from life as he lived it. Certainly, his adult libertarianism fits snugly with an upbringing of family, work, and faith. But millions of Americans have had similar traditional upbringings in small towns and haven't ended up with a firm libertarianism.

Paul's adult political thought started to become something developed, not merely instinctual, when he read *Dr. Zhivago* in medical school. An NPR reporter found something deep in Paul's love for the book: "You can see how the book would resonate: It's about a young, idealistic doctor who wants nothing to do with politics until he is co-opted into the Russian Revolution." Paul's own comment to the reporter was more matter-of-fact: "I was intrigued with this whole idea of Communism and the terrible conditions in Russia and the way Zhivago had to put up with this."

He also began reading the works of the Austrian economists in medical school, starting with Hayek's famous *The Road to Serfdom* (1944), the book that made Hayek's popular reputation. It argued the Western world was dabbling dangerously with the same economic planning and controls that defined their fascist enemies during World War II. Paul moved on to the economic and politi-

cal texts printed or promoted by the Foundation for Economic
Education (FEE). That group, in those dark New Frontier days,
was one of the few sources for the Real Stuff, the serious Old
Right libertarianism uncontaminated by the Buckley/Goldwater
bellicosity against the Soviets.

Paul moved on to some other foundational works of the modern libertarian tradition, including the controversial Russian
émigré novelist and philosopher Ayn Rand. Paul embraced this
writer, whom all the proper people condemned as hideous both
in prose and in ethics. Paul saw that she became a word-of-mouth
sensation (which she remains to this day, her novels still in print
after many decades and still selling as well as new novels) "because she was telling the truth and people were eager to hear it."
He later noted his disinterest in her atheism, but he continued to
recommend her for those wanting to understand his perspective.

Paul graduated from Duke in 1961 and did his medical in-
ternship at the Henry Ford Hospital in Detroit. He developed a
rosy-eyed view of the way medicine could work, and did work,
in the days before Medicare and managed care and the complicated medical-insurance complex of today. (While Paul is not the
type to get number-wonky when he reminisces about the good
old days of direct doctor-patient medicine, National Bureau of
Economic Research economist Amy Finkelstein did a 2011 study
finding that the widespread use of Medicare insurance accounted
for 40 percent of increased medical expenses from 1950 to 1990.)

During the Cuban Missile Crisis in 1962, young Dr. Paul got
a draft notice. He talks about his military service on the campaign trail sometimes, though not as often as you might think.
He always mentions he was drafted. He didn't want to be in the
army, so volunteered to join the air force as a first lieutenant and
practice medicine. Still, he will always point out the draft notice
part, his honesty outweighing cheap electoral advantage. He does
that "because it's true," he says, though he also finds that the vet-

erans of the World War II, Korea, and Vietnam eras with whom he's dealt as a congressman were almost always drafted; they get it. "It's a little bit entertaining" for them, he says, to tell the story of how "I served five years, and how as soon as I was drafted I immediately volunteered!"

He was assigned to Kelly Air Force Base near San Antonio, and the Paul family's Texas years began. He became a flight surgeon, taking care of the medical needs of air force pilots. He worked with a Military Air Transport System (MATS; a now-defunct division of the air force) unit. "It wasn't a fighter group and it wasn't a Strategic Air Command group, but was for carrying matériel around the world. There was also a security element—the general in charge of security for the whole air force" would often travel with him. Paul went to Korea, Japan, Europe, Pakistan, Iran, and Turkey.

"The most fascinating geographic experience I had was, we had a base in Peshawar, very close to the Khyber Pass. You could get a special car and drive up to the Khyber Pass and go into this cave—I wonder if the cave still exists. It was like an underground Hong Kong, like a huge department store where east meets west, Russian products and American products. Some Americans were fascinated with Russian products, their vodka. Aboveground there was fighting and killing but underground there was always one place trading was going on." Paul remembers a commander taking him and some other men up there, pointing out mountain tribes and noting that they'd lived there for hundreds of years, and many have tried to conquer them: "so many outsiders, and no one ever has. That stuck with me."

Paul doesn't sound filled with woe when he talks about his air force service—Paul sounds pretty ploddingly matter-of-fact about everything—but does say that "the last thing I want people to do is *thank me for my service*—that annoys me. I did what I thought I had to do." He remembers one strongly antiwar doctor in the service who rebelled "and he might have gone to prison. I was clearly not anxious to go to prison."

In 1965 Paul was transferred to the Pennsylvania Air National Guard, and he began working and studying in obstetrics and gynecology. He dipped into research medicine and published articles in academic medical journals on toxemia in pregnant women.

During his obstetrics residency in Pittsburgh, he had the abortion experience that defined his attitude toward the practice, a story he sometimes tells on the campaign trail. He even entered it in the *Congressional Record* in his first stint in Congress: "I did not always draw rather stringent lines on abortion until I was forced as a young physician to face up to the problem. I was called to assist one day as many young residents are in an operation performed by a staff member. It turned out to be a hysterotomy, a type of caesarian section with the removal of a 2-pound infant that cried and breathed. The infant was put in the trash and left to die. . . . We as physicians can now save many infants that are born weighing 2 pounds. . . . [F]ollowing this experience I reconsidered my position of 'necessary abortion' and came up with an entirely different perspective." That's another example of the Paul style—in most circumstances even his strong emotions are delivered calmly, measuredly, dispassionately. When I hear him personally deliver a version of this story, it is no more heated than this reads on the page.

After his residency ended, the Pauls wanted to return to Texas, which they'd grown to love during his stint at Kelly Air Force Base. Paul had a potential job in academic medicine lined up with the University of Texas Health Science Center in San Antonio. He learned that the only ob-gyn in Brazoria County was about to abandon his practice. Paul met with him and agreed to take over the office, equipment, patients, debt and all. As of July 1968, Dr. Ron Paul became the ob-gyn for Brazoria County, settling in the city of Lake Jackson, originally a town designed by the Dow Chemical Company in the 1940s.

In 1974 he took on a partner, Dr. Jack Pruett, to help share the load—he'd been dealing with fifty births a month. Pruett had to agree with two principles of Paul's: they'd do no abortions, and they'd accept no Medicare or Medicaid. They'd help arrange adoption when needed, in an adoption market less regulated than today. If an indigent patient couldn't pay a full bill, they'd take care of her anyway. (In those pre-corporate medical days, Drs. Paul and Pruett didn't even have malpractice insurance for years.)

Paul loved the doctor/patient part of his profession but grew weary with the complicated tangling of government, corporation, and medicine that has grown over the past forty-five years. He recognized the monopolistic power that the medical industry's regulations gave him. In an interview for the Kaiser Family Foundation in 2007, he noted that "there are some things that a nurse can do every bit as well as I can do, but by law, they can't do it." Dr. Paul does not approve.

Still, medicine was a good life; Paul had a practice he loved in a place he loved and a growing family, with four sons and a daughter. He could have had a successful, prosperous, and content life as a small-town Texas baby doctor.

Then Richard Nixon happened. Nixon's effects on America were complicated and long-lasting, but one of the little-noted ways Nixon shaped the future of politics is that he turned Ron Paul, Texas obstetrician, into Ron Paul, congressman.

Paul had already been reading his Hayek and his Mises. He was appalled that an American president—a Republican no less, a member of the party of Goldwater—would try to curb inflation (which was caused by Federal Reserve policy to begin with) through wage and price controls. Wage and price controls had failed to work from the ancient Greeks through Diocletian and colonial America and the French revolution, and Paul understood that the inexorable laws of supply and demand dictated that they wouldn't work now, either.

Not only did Nixon violate the principles of justice and free markets through wage and price controls, but on the same horrible day, the day that Paul believes set in motion every economic problem we've had since, Nixon cut the dollar off entirely from its link to gold. Paul had been studying Austrian School popularizer Henry Hazlitt for years, and knew from him that the jerry-rigged solution to the problems of stabilizing international currency markets—a solution forged at the Bretton Woods Monetary Conference in 1944—was bound to fail; allowing foreign governments to exchange dollars for gold at a fixed rate and not keeping a firmer control of the money supply would guarantee that particular international monetary system could not long survive.

Another man shocked into action by Nixon's action that August day was David Nolan, who responded not just by thinking about running for office himself, but by creating a new political party, the Libertarian Party. Its story will intersect Paul's later.

Paul was getting deeper into the worlds of gold and Austrian economics. He met Mises's student Hans Sennholz at a conference dedicated to promoting the re-legalization of private gold ownership (banned by FDR in 1933) in the early 1970s. "I got to know him very well, and he helped me out with my education foundation later," Paul says. In his last national lecture tour, the grand old man of Austrian economics, Mises himself, went to the University of Houston. "I found the one other doctor in town who had heard of Mises," Paul says. "We closed our offices and took a drive to listen to Mises in Houston."

Paul knew that a disconnect of the dollar from gold would "usher in a new age of rampant inflation and big government and I wanted to speak out. I didn't expect a political career. It was more or less a podium to express myself and let other people know a different way of doing things. That's how I got involved in the campaign in '74. No one was interested in running as a

Republican in '74"—it was the year of Watergate. Paul did not win that first race. He ran on the slogan "Freedom, Honesty, and Sound Money."

"But I became known in the district, and it set the stage for what happened in '76," when the veteran Democratic congressman who had beaten Paul in '74, Robert Casey, resigned. He had been appointed to the Federal Maritime Commission in the Ford administration and a special election was held in April to replace him. The local party knew Paul by then, and he was already familiar to thousands of families through his obstetrics practice. This time he won.

The first time around, Paul did essentially no fundraising. "I remember sending out one letter to a local Republican list, which was very small. It was an amateurish letter and I only got two or three responses. But by '76 there was a breakthrough there. The Republican Party was supporting me and the National Republican Congressional Committee did a fundraising letter for me, got a congressman to sign a letter for me."

Ron Paul was not yet Ron Paul. "They saw me as a typical conservative, and though my beliefs have changed a little since then, not long ago I looked at some of those old letters, and even in '74 and '76 I was always talking about monetary issues and the gold standard. That was never something that became a negative in the district." He was known as Dr. Paul—friends and foes of Paul have both fingered his greatest political asset as the fact that he has delivered four thousand babies in and around Lake Jackson and Brazoria County. His Democratic opponent Bob Gammage complained to the *Houston Post*, "he was portrayed as Marcus Welby and I was cancer." His campaign slogan that second time was "Put Big Government on a Diet." Dr. Paul was going to slim down a fat Uncle Sam. "If I thought Uncle Sam was too fat in '76, you know what I think now," Paul says.

Paul had help from the king of right-wing direct mail, Richard Viguerie. Not everyone understood then how important direct mail was, Viguerie recalls. Everyone was still enamored of TV ads and billboards. But the direct communication via mail to Republican voters was then and now the best bang for the buck for candidates, Viguerie insists, and it worked for Paul. Paul "was not someone carrying the libertarian banner in the way he is these days," Viguerie says. "We sold him as a traditional constitutionalist conservative."

So Dr. Paul went to Congress. Self-consciously an outsider, Paul liked, and still likes, to say that he found prior congressional experience a minus, not a plus, when deciding whom to hire. At the start, though, he did take on a married couple who had previously worked for Steve Symms, Republican from Idaho, a rare pre-Paul example of a libertarian-leaning legislator. Symms had offered bills to eliminate the postal monopoly and legalize gold ownership in his own first term, and entered articles by F. A. Harper, founder of the libertarian Institute for Humane Studies, in the *Congressional Record*. (Symms's hard-core libertarianism softened over the years, and he became mostly indistinguishable from other Republicans.)

The Koopmans were a couple born from conservative politics, meeting through their activism in Young Americans for Freedom. They both went to work for Symms, who used to sign his letters "yours for a free society." Symms, as Roger Koopman recalls, "was keenly interested in Ron's success. He would talk about Ron a lot in the office."

Symms introduced Koopman and his wife, Ann, to his new colleague, and they went to work for Ron Paul. Roger helped Paul assemble the rest of the staff. "He wanted a staff philosophically lined up with him, people there for all the right reasons," Roger remembers. "Not Hill rats or those enamored by the power and prestige of Washington, D.C. He was there to make a difference and advance the ideas of freedom."

Paul told them he explicitly did not want people who would feel their job was their life. Ann was pregnant when they were hired; "Ron was used to being around pregnant women and was perfectly comfortable with me waddling around the office. I was probably the only visibly pregnant person on Capitol Hill. He used to tease me. He knew the baby was due in the fall, and he'd say, 'Let's just wheel you down the hall and I'll deliver your baby—we'll make the Paul Harvey news!' He was a big fan of Paul Harvey. Years later they met and had lunch, which Ron considered a big thrill. But the House was in recess when the day came, so we never got to execute that plan.

"He told us he never expected to find people like Roger and me on Capitol Hill," Ann says. "He thought he'd have to do all his own writing. He didn't think anyone understood how he thought or wanted things said. He was delighted to have us on board."

Ann handled constituent mail. "One letter I had to answer was from a prominent wealthy woman from the Houston area asking that he support government funding of the performing arts. He told me to tell her that his personal preference for entertainment was going to a baseball game, for which he paid the price by buying a ticket. He suggested that patrons of the arts should finance concerts and the like through private donations and fundraising, and that it wasn't the proper role of government to subsidize either enterprise. He wasn't intimidated at all, although she was a powerful and wealthy constituent."

Ann was asked to inform local business interests that in the congressman's mind, "big government, big labor, and big business were a triumvirate that feathered one another's nests and that he was not for corporate welfare or any breaks for big business. I think some people were surprised. He was a Republican replacing a Democrat, and they expected him to be helpful. But if business was coming to government looking for something, he wasn't going to carry their water."

To ensure a staff that understood where he was coming from, Paul relied on people with some history with the original libertarian think tank, the Foundation for Economic Education (FEE), including John Robbins, Bruce Bartlett, and the controversial Gary North.

North was an advocate of the Christian sect known as Reconstructionism—a biblical literalist gang founded by North's father-in-law, R. J. Rushdoony. Their reading of biblical law leads them to promote a mix of economic hyperlibertarianism and social intolerance so severe that the death penalty was prescribed for homosexuality and adultery.

Bruce Bartlett, a young right-winger who was reaching a dead end in graduate school, came to Ron Paul looking for a job. He had read a *Washington Post* article about the special election outcome, "and I remember it said the winner, Ron Paul, was to the right of Barry Goldwater, which in those days sounded good to me. I recently looked at the article again in the *Post* archives and found that it was Bob Gammage [his Democratic opponent] who said Paul was to the right of Goldwater, but I just remembered the 'to the right of Goldwater' part."

Bartlett had published a few articles in FEE's journal the *Freeman*, which helped secure him the interview in Paul's office. "There was a little bookcase in his office completely filled with every book published by FEE, so I lucked out there. I was hired as a legislative assistant.

"I'd spend time going to committee hearings, in case Ron came in and wanted ideas for questions," Bartlett says. "We were always looking for opportunities to pursue Ron's agenda, and one thing he always liked to do was, whenever a committee passed out a bill, he liked to have a dissenting view attached to the committee report. The main thing I used to do was write those dissents and try to get other Republicans to sign them. Then [Federal Reserve chairman at the time] Arthur Burns would come in and

John [Robbins], Gary, and I would contribute ideas for on-the-record questions Ron could ask him. Every Fed chief since 1976 has dreaded the time for Ron to ask questions."

Paul recalls his first vote as being on some educational program that the party expected him to vote for. And he didn't think he should. But he was new, and "I was hounded by my colleagues, this was considered the conservative thing to do, this is a very important vote. . . . They pounded and pounded away at me. And I thought, 'I guess I better do it.' I didn't know what I was doing. It was my first day on the floor. And I switched my vote."

It's like a superhero origin story, this moment of fateful choice in the life of Ron Paul, like the horrible mistake Spider-Man makes before learning that with great power comes great responsibility. "I thought about what I had come here for and decided no, I can't vote for that. I'm not going to respond to them hounding me. So I put my card in to change the vote, and it was too late; I had to go to the floor and announce the change on the record. But I actually switched my vote! That was the last time I was tempted to be pressured by anybody to cave. If you cave once, people will be pestering you for votes forever. That was the last time I even thought of compromising my vote. I slept better that night."

After that first slip, he built a reputation that allows him to avoid niggling pressure, both from his colleagues and lobbyists. Everyone began to get it: no point in trying to logroll or shift Paul from whatever vote he thinks is right.

Paul had already begun earning the nickname of "Dr. No." He doesn't like it. It's too negative, he'll say. He stands *for* something—American liberty. Bartlett recalls Paul was "never happier than when the vote was 434–1. It was his way of making a point."

Paul kept his family in Texas and went back to them every weekend. Rather than buy or even rent a home in Washington, he crashed with an aunt living in Virginia. Paul didn't think he was there to settle in. He saw being a congressman as a teaching opportunity for small-government principles, Roger Koopman says, especially when he violated expectations for a freshman congressman.

"He was getting advice all the time from his Republican colleagues to be an advocate for the interests of his district. That's the way you get reelected. Ron almost looked forward to these opportunities to vote correctly, whether it was something that enriched his district or not. A good example was the Johnson Space Center. He absolutely mortified many Republicans that he'd vote against appropriations for the Johnson Space Center, since a lot of people in his district were employed there, and vendors and so forth. He was told he could never get reelected taking those sorts of stands." After that first special election victory, Paul's constituents had little sense of what sort of man they had sent to represent them. (Except that they knew he was a doctor. Bartlett also dealt with some constituent mail; he remembers one woman who was thrilled to discover that her new congressman was an ob-gyn and sent him "a long letter full of several pages of female plumbing problems.")

It's not that he was uncollegial, with his never staying in D.C. and his "Dr. No" reputation. But he'd be collegial in his own way. Rather than doing the cocktail party circuit, he joined a congressional baseball team (Republicans versus Democrats). "As far as his interactions with colleagues, and I think this has been true throughout his career, he was well liked, though he may have been looked at by some as quaint and unrealistic and naïve," Roger Koopman says. "But for a person to hold such strong, unshakable principles and at the same time to be as professional and gracious as Ron gained him a lot of respect. I didn't sense any animosity toward him, though he was a lone wolf who cut his own path."

Paul was not nearly the nuisance even to his own Republican

colleagues that he would later be seen as. "In those days the Republicans had a very small membership," Bartlett recalls. "They had only around one hundred and forty seats in the House and so were essentially irrelevant. That gave Ron a lot of freedom to do whatever he damn well felt like. It didn't matter, and the Republican Party in general didn't matter. There was nothing to do but make a stink as best you could." Paul was more than capable of doing that on his own. Though he craved ideologically simpatico staffers, Bartlett notes that "Ron is not somebody you had to write out speeches for. He knew the issues as well as we did."

Still, Paul didn't love his colleagues as much as he might have, that first go-round. "Privately he was very disappointed in many of these people," Roger Koopman says. "He went up believing he'd be locking arms with genuine freedom fighters, and when he got there . . . I don't know if he felt that some of his colleagues had been worn down by the fight and jaded or whether over time they'd just learned to be more political and choose their battles, but he did conclude that a fierce warrior spirit he expected to see there was largely lacking."

Specifically, Roger recalls, Paul was surprised at how pusillanimous his Republican colleagues were about a bill regarding the District of Columbia gun ban (which was overturned decades later in the 2008 Supreme Court case *D.C. v. Heller*, the subject of my previous book *Gun Control on Trial*). Paul was already close with Larry Pratt of the Gun Owners of America, and Paul was amazed to find he was more or less alone on this issue. "Even the NRA had faded into the wallpaper and they were not going to take on the issue at all," Koopman says. "It ended up that Ron stood alone in fighting the D.C. gun ban."

He failed to halt that Second Amendment–shredding law, and Roger remembers that "that was an eye opener for him. He recognized from that point forward that he had a special role to play, not because he sought distinction or glory or wanted to claim any credit, but because he realized there would be times when he was the only one who would have the courage to fight.

"That's why he made so many special order speeches in later years—to get things on the record, for history, these ideas he thought had to be expressed but no one else in Congress would say," Koopman says. "Somebody else could have gotten discouraged and gone back to Texas and never come back. But he wanted to make sure someone was standing for these ideas in Washington."

Sometimes Paul had to stand against his own staff. "If you were a conservative, you were thought to be pro-defense, and if pro-defense, you were thought to be pro the B-1 Bomber," Roger Koopman says. "I unthinkingly went along with that. Some others understood Ron might not be reliable on the B-1 Bomber, which was a good assumption, and they were feeding me materials and information to pass on to Ron. I got Ron's attention. We had a small staff; he did not use the full allocation of tax money to have a full staff. There was lots of unspent money, and I really respected that. But I'd pass this info on to Ron. Occasionally I'd try to discuss the B-1 with him and he realized that I was feeding him information I was getting in order to get him to vote in support of the B-1 bomber.

"He was always cheerful, never an unkind thing said, never used bad words, never talked off-color. It was just his nature. But he could still be very firm when he had to be. We were walking down the hall, and he said to me, 'I don't like my staff lobbying me.' And I understood later that he was right about the B-1 bomber. It was a boondoggle and no conservative should have been supporting it."

Right-wing direct mail maven Richard Viguerie remembers with a sort of head-shaking admiration that Representative Bob Dornan of California was one of the national conservative illuminati who signed a direct mail letter in support of Paul in his

1978 race that returned him to Congress. Dornan was popularly known as "B-1 Bob," since getting the government to buy more of the unnecessary, overpriced, and underperforming bomber jets was a high political priority for him. "Ron wins, gets back to the House in January of 1979, and his first vote is against the B-1 bomber." Nothing would sway him on this; his chief of staff in the early 1980s, Lew Rockwell, recalls phone calls from President Reagan attempting to win Paul over for the team on B-1 funding, "and it had no more effect on him than any lobbyist calling him. I mean, he gave him the respect he felt the president was due, but it had zero effect."

Paul was fortunate to quickly get the committee assignment he most wanted. "My motivation in going to Congress was monetary policy," Paul says, "so my only request was to get on the banking committee, as it was called at that time, and I got on the domestic monetary policy subcommittee and I'm now [as of 2010] chairman of that. I started a lot of correspondence with newsletter writers into hard money investment, and they'd do lobbying for certain bills when I'd hear of some terrible bill. It was slower than the Internet but it did work. I'd send out messages to them, they'd write them up in their newsletters."

Paul set himself up in opposition right away. He was the lone banking committee member opposing a bill to refinance the International Monetary Fund (IMF). To Paul, the IMF was an unconstitutional mess of Bretton Woods silliness—it was conceived at the conference—and international banker manipulation that benefited elites both in banking systems here and in governments abroad.

Gary North wrote the minority report making Paul's points against the IMF. "That minority report," North later remembered, "so completely amazed the bipartisan establishment that Ron Paul was invited to testify to the Senate Banking Committee on his reasons for opposing the funding. I had never heard of this before: a freshman congressman invited to share his views with a Senate committee. I have not heard of it since."

Having won a special election, Paul had to run again mere months later for the regular November 1976 contest—which he lost. "It was very, very close and we contested it. We considered it an absolutely fraudulent vote. There were a large number of illegal votes, and we could have proven it. People were voting with addresses that were gas stations and parking lots, addresses where no one was living. We did work on that and we had evidence.

"I was convinced I was cheated. It's a terrible system; I thought you just needed to have a good candidate, have good beliefs, raise some money, run a good campaign. But one thing I didn't know about was ballot security, and we didn't have that in that election. I figured if I'm serious I ought to do it all the way."

Paul challenged the results in state court, but his challenge was tossed and Democrat Bob Gammage's victory ratified. Ron Paul has since become famous for aggressively good constituent services, helping solve any problems the people of his district might have with Social Security, the IRS, and immigration. Bartlett recalls that Paul's attitude toward that sort of service got more serious after he lost an election by just a few hundred votes.

Still, at the time Paul told the *Houston Post* that he didn't need the job, and was already speculating about starting an educational foundation or some other means of propounding his philosophy—which the *Post* already noted combined opposition to welfare for both the poor and corporations. Paul told them intervention in foreign countries was "reprehensible" and support for the Chilean dictatorship a "monstrosity" and that he leaned toward amnesty for draft evaders.

His narrow, suspicious loss pushed Paul to run again against Gammage, the man he was sure had cheated him, in 1978. "In '78 I personally knocked on ten thousand doors; the family helped and I had pretty good volunteers," Paul says. "The money didn't come in like it does now, but I raised enough to get on. I didn't have a national reputation yet, but I had a three-thousand-name list that the national party sent. Those kinds of lists kept building."

Ron Paul began his second stint in Congress, but being an obstetrician was still important to him. He would tend to fly back home Friday morning and deal with patients, and do the same on Saturday. He had to mostly start fresh with staff, though John Robbins continued to work for him. The Koopmans and Gary North had left Congress; Bruce Bartlett had gone to work for Jack Kemp, and continued a career path as an economic analyst for other Republican politicians and right-wing and libertarian organizations before deciding in the second Bush administration that conservatism had become maddened by a tax-cuts-over-all mentality that no longer makes economic or fiscal sense for a government grotesquely in debt. Bartlett is hostile to almost all Republicans now but still has kind words for Ron Paul. "Ron was a very easy person to get along with," Bartlett says. "And last time I saw him, he seemed the same identical person, hardly aged at all, and his views exactly the same. He's very humble, never puts on airs or acts like he's a big shot, and that obviously is a very endearing quality and one of the reasons he's managed to get elected so many times."

In his second stint in Congress Paul picked up as chief of staff Lew Rockwell, who had been an editor at right-wing publisher Arlington House and had edited a journal called *Private Practice*, which promoted free-market medicine. Paul was still an obscure figure on the national scene, as nearly all congressmen are obscure. But those keenly attuned to small-government tremors were already detecting him. One of the few popular expressions of small-government sentiment in the immediately pre-Reagan era, those early flashes of the mass mistrust of post–Great Society liberal governance that tore down President Jimmy Carter and elevated Reagan in 1980, came from William Simon, who had been Treasury secretary under Nixon and Ford, in his bestseller, *A Time for Truth.* (The book was ghostwritten by former Ayn Rand associate Edith Efron.) Simon had already singled out Ron in 1978 as "an exception to the Gang of 535."

Paul continued to stand for things that few other Republicans—in some cases few other congressmen at all—stood for. It was the late Carter era, and the government was contemplating bringing back the military draft. (Draft *registration* did resume in 1980 under Carter.) Paul testified against the idea in April 1979, in language that seemed designed to appeal to the right: "A draft will not substitute for, nor create, a morally and spiritually strong nation. Force cannot overcome apathy."

Paul also became, often with the help of his Democratic colleague from Georgia, Larry McDonald (chieftain of the John Birch Society—yes, he was a Democrat), a prominent voice defending what he saw as American sovereignty against the threat of foreign entanglements, military and otherwise. Paul's tendency to talk about these sovereignty issues is one of the things that makes some libertarians uncomfortable with him, seeing him as too entrenched in the concerns of the populist right. (Roughly, that foreigners, in international organizations both obvious and arcane, and on the other side of our borders, are threatening our way of life. It's a viewpoint Paul has had more sympathy for in his career than most libertarians have, though it's one he has hardly stressed in his past two presidential runs.)

Paul loved to lecture his colleagues on the evils of the United Nations, the IMF, the World Bank, and later, in the 1990s, the World Trade Organization, which "usurps our rights and our privileges and interferes in our legislative process, especially in the area of environmentalism and labor law."

Even in the age of Ford, Ron Paul supported Ronald Reagan, and in 1976 he led the pro-Reagan Texas delegation to the Republican National Convention. Paul was one of only four congressmen bucking the establishment and the incumbent, Ford, to support Reagan that year. Still, in the 1980 contest, Paul was first attracted to Phil Crane, "a hard-money person; we got along real well." When Crane dropped out, Paul returned to Reagan. And quickly regretted it.

"I had always liked Reagan. He liked the gold standard and was a really nice person to visit with," Paul says. "But his first budget was terrible; I was the only Republican who voted against it. I thought he was terrible when it came to changing the course of history. There was a little tax cut here, then a little tax increase here. The deficit exploded. He wasn't for a noninterventionist foreign policy. I thought government would shrink under Reagan and it exploded like crazy."

Paul was out of step with the general tenor of conservative fiscal thought during the Reagan era. "I disagree with the supply-side argument that government debt doesn't matter," Paul wrote later. "Debt does matter, especially to future generations that will be asked to pay for the extravagance. . . . [R]esponsible people restrain their borrowing because they will have to pay the money back." Supply-siders just thought cutting taxes would take care of itself in increased growth. Paul still regrets what Reagan ended up doing to the standard Republican ideas about fiscal probity. "Now you have Dick Cheney saying that 'Reagan taught us that deficits don't matter.' I never agreed with that, that deficits didn't matter that much. They always matter."

Paul's 1980 victory over a Democratic former assistant district attorney, Mike Andrews (who made it to Congress himself later in a different Texas district), was narrow, fewer than six thousand votes. Paul was far from the unstoppable electoral force he's become during his current stint in Congress. During this second full term, Paul guaranteed his future as an American political icon, cementing his reputation as *the* hard-money man in Congress.

Paul recalls rallying the hard-money newsletter world to try to defeat that IMF bailout bill he opposed as early as 1976. But it was his work on the Gold Commission that ensured him his permanent place on the honor roll of those who believed that gold was as good as gold, and that government monetary policy in a post-Nixon age was a slow-motion train wreck destroying the value of the dollar, Americans' savings, and everything good and decent about America.

From this hard-money world came a wave of popular books and newsletters and TV shows telling people how they could weather the coming enormous mess of devaluation, crisis, and collapse that government policies were all but guaranteeing. Hard money, foreign currencies, food, weapons, property in small towns or in the woods—serious problems were brewing on the horizon, dark and grim but with a golden lining, and Ron Paul was one of the few people in political power who thought he saw that coming.

Even before the current age of Paul, and far beneath the academic empyrean of the likes of Mises and Hayek, more populist sources of economic education were already pushing a pop Austrianism—including self-created educational foundations outside the traditional academic world and newsletter writers and pamphleteers who see and sell doom, depression, and an always-coming collapse.

These pamphleteers and enthusiasts and snake-oil salesmen play into some classic American presumptions, anxieties, and points of pride. One such presumption is that the powers that be are almost sinisterly unable to manage currency in a way that redounds to the benefit of the populace at large—meaning, big banks and financial institutions are in bed with the government and laughing at the middle and lower classes (or using the lower classes as a weapon against the middle). Another is that a keenly applied frontiersman mentality can help you and your family (or, you know, just you) survive and thrive even if the centers of urban civilization are falling apart—city-slicker life might be getting ugly, but you can make it on the interstices, if you are clever and resourceful enough. And save your gold and silver. The creators and consumers of these hard-money and crisis books and magazines found a political hero in Paul because of his role on the Gold Commission.

Paul earnestly wanted to be on this commission, created by act of Congress in 1980 but not fully constituted until 1981. "I did so little phone calling to say 'can you do this for me,' because it was always tit for tat," Paul says. "'You do this, I do something for you.' But I do remember when I heard the Gold Commission was actually being formed, Kemp was sympathetic to gold, and I called him and asked if he would help [make sure Paul was on the commission]."

Paul had launched his own nonprofit educational foundation on the FEE model in the 1970s, called the Foundation for Rational Economics and Education (FREE); he issued a newsletter called *Ron Paul's Freedom Report*. Back in Congress, Paul began throwing educational dinners for administration officials on the subject of gold; the new president, Ronald Reagan, was theoretically in favor of reversing the economic sin of Nixon's that had led Paul to Congress: severing the U.S. dollar from any gold backing. "Ronald Reagan told me that no great nation stayed great once they got off the gold standard," Paul says. "But the rest of the administration was hostile. [Treasury secretary and former Wall Street leader] Don Regan didn't even want to have open hearings on [a return to gold] because it might 'interfere with the market.'"

Paul's ally in goldbuggery during this era was Senator Jesse Helms of North Carolina, who spearheaded the effort to make it legal for private contracts in the United States to be denominated in gold again, more than forty years after Roosevelt had barred U.S. citizens from even owning the metal. Private ownership had been re-legalized in 1975, and Paul credits his pal Jim Blanchard of the National Committee to Legalize Gold for leading the political charge on that. Paul likes to tell a story about his meeting with Federal Reserve chair Paul Volcker in this era, and noting that even this chief of the paper money system, during their breakfast meeting, insisted that one of his aides tell him what the price of gold was that morning the second he walked into the room. Even he knew, Paul says, that gold gives signals about the strength of the dollar that the government shouldn't ignore.

Ron Paul comes from curious corners of American intellectual life that few understand; he's the cult politician par excellence, in the sense that his enthusiasms tend to be mightily, but thinly, held across the American political landscape. When you get what Paul gets, when you are into what Paul is into, you tend to be really into it, with the fervor of the lover of a cult movie or cult band. The world of "goldbugs" is alien to most Americans, happy to just spend their paper notes and blame inflation vaguely on greedy unions or corporations or "supply shocks" or anything other than a Federal Reserve that chooses to make more dollars to help the government spend beyond its means. Having more of those dollars around, as Paul and Paul's fans understand, makes the ones in your pocket worth less. That is inflation, and Ron Paul, who remembers his old German grandmother obsessed with getting land for their money because she remembers the time money became worthless, thinks it's nothing more than theft.

Paul didn't just want to link the dollar to gold or back it partially by gold. He wanted to completely eliminate the power of the government to manipulate the stock of money. He wanted the dollar to be, by definition, a certain weight of gold. From that point on, the market could sort it out.

There are many complicated and sophisticated arguments about how well a modern industrial economy of the sort that we've developed on fiat money could work if money were once again gold; the most influential one seems to be that the discipline of gold limits the ability to create new credit, the new credit that gooses the economy and supposedly creates growth.

The Keynesian/inflationist trick, at its most basic and human level, is this: the real stuff of human wealth comes from human effort. Everyone understands that money itself is not wealth—at least if pressed. We can't eat it or be entertained by it or educate ourselves with it or shelter ourselves with it or have an adventure with it in and of itself. But when money is working well, as that thing for which everyone else is willing to exchange anything because we know that everyone else is willing to exchange everything for it (at some ratio we think we roughly understand through prices), then money is a valuable tool we can use to get all those things we want.

The real wealth of the world, either goods or services or shelter or even land (unimproved land essentially meets the needs only of someone who wants to hang out in a meadow or forest), arises from human beings and their skills, knowledge, and physical power; people have to *do something*, to transform the physical stuff of the world. The food gets in the stores because someone grew it, harvested it, processed it, transported it, and put it on accessible display in a place arranged where people know they can come and get food.

The real trick of a functioning economy is getting people to expend that effort. The communists thought they could do it by force, and then after the state would supposedly wither away, by a desire to just help everyone out, knowing everyone else was working to help you out. Great things can happen because of the desire to just help people out—the world of Burning Man is a great example of this—but it works at Burning Man because everyone there is in a position of abundance to start with. The economic problem of man on earth is that we don't have all the abundance we want without effort.

In our society, money encourages people to make effort. They know they can exchange money for all the goods and services they want. That makes it better than just food, or just shelter. Payment in cash rather than in kind increases our autonomy.

So, say the inflationists, although it is true that an increase in money does not by one whit increase production or wealth in and of itself, it motivates human beings to get up off their asses, organize, and increase the production of things and services that are actual wealth. You can also pay people to do things, such as, in the classic Keynesian example, dig holes and fill them up, or in the more modern variant, build bombs and blow people and things up, but neither of those activities adds to real wealth because they don't make things better for anyone.

But here's the Keynesian/inflationist trick: Paying people to do those useless or destructive things puts money-tokens in their hand. And those money-tokens represent what they like to call "demand." Demand in the form of hole diggers or munitions workers with cash in their pocket can encourage the people actually making useful things to make more of them to get that money so they can spend it to get the things they want, sending that money circulating through the economy and creating the "multiplier"—each dollar when it changes hands represents more demand for something that encourages the people doing useful things to do more of them so they can get the cash and spend it, and a virtuous cycle ensues.

Sounds great, doesn't it? But the money-tokens exchange for the things people want at a certain ratio. That ratio, by the laws of supply and demand, shifts depending on how many of the money-tokens are flying around from hand to hand. The inflation of the money supply leads inevitably to the inflation of prices, which means savings are worth less, which harms people's ability to make plans and preserve what they've already earned.

We've seen many examples of hyperinflation in which government's attempts to keep goosing the economy with more trading tokens reduces the value of all of them to practically nothing. And when those tokens are all most people have, the social chaos and damage is severe. In Paul's reading of history and economics, governments are not very good at reining in inflation once they let it loose.

And the only sure way, Paul thinks, to make sure governments don't walk that inflationary path—as the ones making the new money, government and its cronies benefit from it first, before its value is diminished—is to pretty much eliminate their ability to make more money. That means to make money a specific thing that is difficult and expensive to make more of, or to bring into the human world in a usable form, like gold, or silver—both of which have maintained their status as things of both use value and near universal exchange value for much of human history.

That's why Ron Paul has that perplexing obsession with gold.

Paul had already introduced a bill that would essentially return the United States to a gold standard, in January 1981. Paul was watching Federal Reserve policy get worse before his eyes. He fought against the Monetary Control Act of 1980, which allowed the Fed to greatly expand the list of things it could "monetize"— create money against the value of—to include foreign securities, and which increased the Fed's regulatory authority over the rest of the U.S. banking system, as well as lowering reserve ratios for the banks the Fed did have authority over, giving them more inflationary power. But he lost.

Paul, gold's best friend in Washington, was on the commission, but he was unwittingly a token as phony as paper money. The fix was in, as the *Washington Post* observed at the time: fourteen of the sixteen members were known to be against any return to a gold dollar. One member, Henry Reuss, a Democrat from Wisconsin, once ostentatiously crumbled a hard-money newsletter and tossed it to the floor during a meeting. Three Federal

Reserve governors were on the commission and unlikely to scuttle their own discretionary authority and power.

Paul got a chance to tell President Reagan, still ostensibly pro-gold, in person that the Gold Commission was inimical to the metal as money. Reagan was sorry to hear that but, once in office, he never put his political heft behind the cause. A recommendation to make the dollar gold again got only two votes from the commission, but they did by a vast majority decide to start coining an official U.S. gold piece again, the Gold Eagle.

Paul helmed the minority report of the commission, published as the book *The Case for Gold* by the libertarian think tank the Cato Institute in 1982 (co-authored by Lewis Lehrman). Anna Schwartz, monetarist Milton Friedman's partner in the magisterial *Monetary History of the United States*, praised the report in the *Journal of Money, Credit, and Banking* as delivering a "generally high level of scholarship on U.S. money and banking history."

The book is both a compact history of American monetary policy and a multifront defense of gold as money, or at least an attack on unbacked government paper as money. Paul points out that living in an environment of constant and unpredictable inflation leads to a world where speculative investments and complicated schemes to just protect the real value of your savings "replaced productive efforts, savings, and planning for the future." Far too much energy and effort is spent, in an inflationary world, on financial shenanigans and speculations necessitated by the shifting value of currencies, both in and of themselves and in relation to each other. Those would not be necessary if all government monies were merely defined weights of gold.

Paul and Lehrman explain the long history that bred mistrust of paper money in America's bones, and enshrined in our 1792 Coinage Act that only gold and silver should be money. Paul stresses throughout that big central banks are good friends to big government, helping them to finance both wars and giveaways to the well-connected without having to resort to immediate taxa-tion. He also writes of government's history of absolving banks of the legal obligation to make good on their promises, such as to redeem their paper for metal, which helped cause the supposed chaos of nineteenth-century, pre–Federal Reserve banking.

Paul also makes the case, following the thinking of his Aus-trian economics mentors, that even in 1819 the United States was already suffering boom-bust cycles based on bank inflation followed by contraction. He also demonstrates how historically deflation need not mean a fall in production or real economic growth, and how it was not the gold standard per se, but the attempt to prop up a gold-exchange standard at an artificially high level, along with inflationary paper creation, that precipi-tated the British and American economic crises in the 1920s and '30s that led both countries to abandon any ties to the classical gold standard.

Paul sums up the heart of his preference for metal over paper: "Gold money is always rejected by those who advocate significant government intervention in the economy. Paper money is a device by which the unpopular programs of government intervention, whether civilian or military, foreign or domestic, can be financed without the tax increases that would surely precipitate massive resistance by the people."

That same year, 1982, Paul won reelection to Congress unopposed.

Paul had become an early and loud voice warning of the dan-gers of U.S. involvement and intervention in the Middle East in the early 1980s as well, and against any expansion in the cost and reach of the U.S. military. He complained from the House floor during the Reagan era about increasing military budgets and the creation of new "rapid deployment forces." In 1981, speaking against the sale of AWACS planes to the Saudis, he said some-thing scarily prescient: "By permanently putting soldiers there [in

Saudi Arabia] we are making a commitment, whether explicitly or not, that we may soon regret."

Paul had become pretty disgusted with the way things went in Washington by the early 1980s. Paul watcher Mike Holmes is from Texas and had known Paul from Paul's days hanging around Libertarian Party events even before he became a politician. Holmes worked with Paul later when Holmes was a major force behind and treasurer for the Republican Liberty Caucus, an organization promoting libertarian ideas and candidates within the Republican Party, launched in 1990, which Paul chaired in the late 1990s. Holmes remembers coincidentally ending up on a plane from D.C. back to Texas with Paul in 1981, and having the congressman earnestly tell him, "Mike, if the American people knew what happened here in Washington every day, there'd be a revolution in the street." David Boaz of the Cato Institute recalls meeting with this most libertarian of congressmen when Cato relocated to D.C. in 1982, and suggesting to Paul that he write more op-eds for major national papers pushing his libertarian line. Boaz remembers Paul saying that he didn't think there was any chance an establishment paper would give him any play, even though other right-wingers such as Jack Kemp were succeeding on op-ed pages. "We thought that was the wrong approach," Boaz says. "It's not that he was completely wrong—certainly Jack Kemp was more popular with liberal elites than Ron Paul would be. But we didn't think it was right for libertarians to have that sort of bunker mentality that no one would want to listen to us."

Paul brought his second stint in Congress to an end by choice, making a failed 1984 run for the GOP nomination for a Senate seat after John Tower—the first Republican senator from Texas in the twentieth century—resigned. Three other Republicans fought it out in the primary with Paul, including fellow congressman Phil Gramm, a former Democrat driven from his party for his fervent support of Reagan. (Gramm had on his team a young Republican operative named Karl Rove.)

This election was the national debut of Ron Paul, the libertarian Republican. He boldly differentiated himself from Republican verities on issues from foreign intervention (getting *New York Times* attention for declaring in a debate that in his world, "We don't interfere, we don't meddle, we don't send the CIA out to murder people") to drug abuse (Paul was already speaking and writing on the futility and silliness of drug prohibition in the era of a phony "just say no" campaign hiding a massive escalation of the government's war on drugs). At the same time, Paul was the National Taxpayers Union's certified "most ardent anti-spender in Congress." Confusing man, this Dr. Paul.

The Senate campaign got tricky and ugly. Those on Paul's mailing list received a letter from the Gramm campaign urging them to go for Gramm. Paul assumed someone with Gramm had stolen the list and thought about pursuing legal action against the Gramm campaign, but did not.

In the end, Paul came in a distant second of four in the primary, garnering only 16 percent to Gramm's 73 percent. Paul wasn't burned up about it, he says. "It was not what I wanted to do. I wanted to practice more medicine, but lots of my supporters thought I should maybe run for Senate. I didn't have high expectations. I knew the odds, but like when I first ran for the House, I did it for reasons other than that my life depended on getting elected to Congress. But I built up more support around the state." That support, held tightly like gold resistant to inflation, would do him good twelve years later when he returned to Congress.

Paul shifted himself to somewhat of a national libertarian gadfly without portfolio. The first trial of a young man for the crime of not registering for the draft was under way in 1985; his name was Paul Jacob and he was a self-conscious libertarian and fan of Ron Paul. He asked that Paul testify on his behalf.

"It wasn't certain he'd be able to come in," Jacob says. "We were trying to figure out a way, get a small plane. I remember someone in his shop asked if we could buy gas, they had to reimburse the private plane. We contacted him four or five times afterwards to get him to send us an invoice. Finally Ron just said he would pick that up.

"He was absolutely fantastic on the stand. They brought up the argument, isn't this a radical view that wants to destroy all government? He started out saying that when he was in Congress he voted against ninety-nine percent of bills, because government is a massive threat to every individual in the country." Jacob had scrawled "smash the state" on some piece of paperwork in evidence; the ex-congressman was confronted with this. "'I wouldn't use that sort of rhetoric, use the term "smash the state," I may not have used that verb,' Ron said, 'but I certainly agree with the sentiment.'"

CHAPTER FOUR

———— ★ ————

RON PAUL, LIBERTARIAN

In 1987, Ron Paul returned to politics. In one way, he was trying to come back bigger than before; in another way, much smaller. He aimed not just for the House or Senate, but the White House. But he chose to do it not as a Republican—a party actually known to win elections here and there—but as a representative of the Libertarian Party.

The Libertarian Party was founded in 1971, and Ron had been showing up at Texas meetings of the party from the beginning. Longtime LP hand Mike Holmes remembers from their earliest meeting "this doctor showing up, a fairly tall, thin guy. He was notable back then among libertarians for looking like he could buy lunch—not that he did! But he did pay for whatever dues were being collected for the newsletter or whatever. We all knew he was very good on the gold issue, and was already referencing himself as an Austrian, very into Mises and knew who Rothbard was." Paul says that even before jumping ship from the Republicans and joining it officially, he figures he'd been to more LP meetings in the 1970s than most actual LP members had been.

Since leaving the House in January 1985 after his failed Senate bid, Paul was again the full-time doctor and family man, keeping FREE going, lecturing nationally on politics, starting a call-in service for weekly mini-lectures on politics and policy, and building his newsletter series to include *Ron Paul Investment Newsletter* and *Ron Paul Survival Report.* That was all for his already substantial national audience of fans, under the national radar. But he made the papers again by publicly resigning from the Republican Party.

In a January 8, 1987, letter to then-GOP chair Frank Fahrenkopf, which triggered an Associated Press story, Paul told his old party to buzz off, and told the world why. How could Paul remain a member of a party that had "given us unprecedented deficits, massive monetary inflation, indiscriminate military spending, an irrational and unconstitutional foreign policy, zooming foreign aid, the exaltation of international banking, and the attack on our personal liberties and privacy"? He was, he said, "wary of the Republican Party's efforts to reduce the size of government," and said Reagan's most significant effect on American politics was that "big government has been legitimized in a way the Democrats could have never accomplished."

It didn't hurt him to leave the party and he wouldn't be back, Paul thought then. He told a libertarian newspaper of the time, *American Libertarian,* that "I had no emotional attachment to the Republican Party. . . . [Leaving] it was a total philosophic movement on my part. I really did come to the point of rejecting the idea that the Republican Party could bring about less government. . . . It was very easy to do and something that I was probably ready to do many years ago."

Paul then gave in to the entreaties of various Libertarian Party insiders and announced he would run for president under its banner. Ron Paul, though still obscure, had become the Ron Paul of today. As *American Libertarian* summed him up when news of the "draft Paul" movement within the LP was first breaking in late 1986, "He is well known to the national news media as an advocate of sound money and increasingly has voiced pure libertarian arguments in opposition to U.S. military intervention abroad and encroachment upon civil liberties by government. He reportedly has an extensive mailing list of past supporters." That's still Ron Paul; the only change is that the list now includes email addresses, not just mailing addresses.

When he started his LP run, Paul assumed the nomination was his for the grabbing. No one else with actual federal legislative experience had ever sought the party's crown. Paul made the very Ron Paul gesture of paying his LP dues in a gold coin, which has tripled in value in dollar terms since. He promised he'd bring in supporters from the constitutionalist right who had either given up on or never cared about the LP.

The Libertarian Party being a gathering of intense eccentrics who treasure the purity of their precious (political) bodily fluids and are above meshing with the wheels of real-world power, at least two prominent Texas-based LP activists opposed their native son Paul's involvement with their party. He had once cast a vote for a harbor project in his district. His run for the GOP Senate nomination in the 1984 race featured, they said, a weaker anti-government framing than a libertarian would like. He had voted against a repeal of the District of Columbia's antisodomy laws—that vote, Paul explained, and not entirely to most LP members' satisfaction, was because the bill also lowered the legal penalties for rape.

Paul seemed proud of association with some very socially conservative types who admired Ron for how true-blue he was on tax, spending, and monetary issues. Even in the LP, some looked at Paul's objection to foreign aid to Israel and saw a dark obsession bordering on or even crossing over into anti-Semitism.

There was one emotional political issue where Paul really was opposed to most LP members: abortion. Paul was, then as now, pro-life. But he was also a constitutionalist and federalist—as with all laws punishing crimes against person and property, he thought enforcing any prohibition against abortion was the business of state and local governments, not the federal one whose office he was seeking.

Some pro-choicers think it's insane and incomprehensible that a so-called libertarian—who is supposed to believe in unrestricted *freedom*, right?—could possibly believe in laws against abortion, whether federal, state, or local. But it makes perfect sense once you grant that pro-lifers *actually believe* that a fetus is a living human the same as a born baby is. At that point, a libertarian believing in laws against abortion makes exactly as much sense as a libertarian believing in laws against murder.

Thrust out of the conservative, religious world into a libertarian one, Paul had to think out loud about abortion, and he revealed early some of the characteristics that make many pro-lifers suspicious of him, even though he's ostensibly one of them. He told *American Libertarian* that despite believing that abortion ought to be against the law, he wasn't comfortable announcing that he believes women who got abortions ought to be thrown in jail.

There's "nothing very attractive to me" about using police power to punish women in that position. Paul got to the heart of what makes him, even as a pro-lifer, inescapably a libertarian: "I despise and detest most of everything which government does. . . . I have not proposed, and I do not know the answers, for what the penalty should be" for abortion.

One of Paul's fiercest partisans in the intra-LP debate was his old pal the libertarian firebrand and omnipresence Murray Rothbard, also the most ferocious living student and disciple of Ludwig von Mises. Rothbard was already a friend and occasional advisor to Paul during his stint in Congress. Rothbard, after a lifetime as a key figure in nearly every libertarian movement institution of any influence, had by the late 1980s alienated himself from most of those institutions, largely through loud and public denunciations of them when he felt they strayed from the proper libertarian line. After such a break with the Cato Institute, the libertarian think tank he helped found in 1977, and with the billionaire libertarian financier Koch brothers, Rothbard in the 1980s became the intellectual linchpin of the economic education group the Ludwig von Mises Institute, founded in 1982 by Paul's former chief of staff Lew Rockwell. Paul has had a close relationship with the institute ever since.

No one hated the state more than Rothbard. Still, he admired Congressman Paul and hyped him to the LP in the 1987 buildup to its Labor Day weekend convention in Seattle. While acknowledging that major party political experience is more of a detriment than a credit, Rothbard assured his fellow libertarians that "I know that Ron Paul has become increasingly radical and libertarian in his years in Congress," that he is that rare, perhaps unique, professional politician who had managed to "gain . . . the support of a mass of the voters on a platform of consistent opposition to special privilege."

Rothbard had long held tight to a strange vision, a vision that he believed was now taking form before his eyes in Ron Paul. He saw potential in Paul as a political figurehead—potential that, to a surprising degree, has been realized, two decades after Rothbard saw it. The rise of an influential mass Paulite movement did not begin in his 1988 LP presidential run, as Rothbard had hoped. But it is happening now. Rothbard had his own political thought forged by the Old Right and believed that the conservative tendency had died only in the political elites, that it remained the slumbering underground belief system of a great mass of Americans disengaged from and disgusted by politics. What has happened with the Ron Paul Revolution since 2007 indicates Rothbard may have been right.

"There are millions of Americans who are instinctive Old Righties, or libertarian populists," Rothbard wrote, "who have been confused and bewildered by decades of Establishment propaganda, and who are there waiting for someone to supply articulation and leadership to resurrect the old cause." Ron Paul, Rothbard believed, was the latest exemplar and leader of this tradition, which kept rising and falling in American history. American revolutionaries, Anti-Federalists, Jeffersonians, and Jacksonians were all real, powerful mass movements of largely libertarian bent. That spirit was still out there. It just needed someone as culturally acceptable to mass America as this humble, serious family man and doctor of conservative mien.

Rothbard feared that the mass of actual Libertarian Party members—he condemned them as "luftmenschen," people of the air, nonbourgeois drifters and hippies who turned off most Americans—were likely to have problems with the staid Dr. Paul. Rothbard was right.

Paul assumed he'd get a stress-free platform from which to advance liberty in the public debate (and help build his audience for his newsletters and mailing list) by hooking himself to this eccentric, underfunded, undermanned party. It wasn't that easy.

What happened on the way to Paul's eventually winning the LP nomination seems ironic in light of the mass right-left, conservative-counterculture coalition Paul has built in the past four years. But Paul in the 1988 context was the candidate of the bourgeoisie, the right end of the libertarian coalition. And that turned off plenty of LP faithful.

Paul faced an unexpected and formidably strange rival from the left side of libertarianism, a man whose antigovernment cred was so strong that he hadn't spent the 1970s *working* for the federal government like Congressman Ron Paul—he had actually been in a firefight with federal agents. He was anti-Western, antigovernment, and at times sounded antirationality itself. And tons of libertarians loved him.

He was Russell Means, the Oglala Sioux Indian activist who had been a leader in the takeover by radical American Indians of the Wounded Knee reservation in South Dakota for several months in 1973. Only in the LP would the four-term congressman have a long, hard fight against a Native American armed rebel who had only evaded jail because of prosecutorial misconduct. "If you just put it down on paper and put Russell Means here and me here and what I've done, you'd think you shouldn't even have to campaign," Paul says. "You'd think I'd get a lot more credibility, but it was a lot tougher than it should have been."

Paul promised the LP he could deliver cash from his existing community of libertarian-leaning conservatives and hard-money activists to produce a well-financed presidential run. Means argued that he could bring in a huge swell of new activists from the political or cultural left, and that the color and drama of his life story—indeed, his very identity—would guarantee media attention that a mere politician running for office such as Ron Paul could not. The LP had enjoyed a multimillion-dollar campaign in 1980, but only by getting around campaign finance restrictions via the clever expedient of picking billionaire libertarian David Koch as their vice presidential candidate. The gambit worked, even with John Anderson stealing the third-party thunder in the press. That year's LP presidential pick, Los Angeles oil industry lawyer Ed Clark, received what remains to this day the LP's best vote performance, with 921,000 votes, more than 1 percent of the total.

But in 1984, with obscure LP stalwart David Bergland as its presidential candidate, the party's fortunes dipped precipitously; the factions associated with the Koch brothers and the Cato Institute had all fled the party. (Paul remembers one meeting with the Koch/Ed Crane faction in that year trying to encourage him to seek the party's nomination; he wasn't interested then.) The LP in the mid-1980s was hemorrhaging money and members. Paul rode in as a potential savior, and groused in an interview with *American Libertarian* that he wasn't thrilled with the notion of having to debate Means. He hoped the LP had enough agreement that such wastes of energy could be avoided. In another sign of that scrupulous Ron Paul fairness that his admirers love, Paul insisted any debate include not only him and Means—clearly the only candidates with an actual chance of winning—but also any other person trying to run, no matter how hopeless the effort.

Means, for his part, tried to win the antiauthority hearts of LP members by stressing such selling points as having removed himself from the Social Security system, not having paid income taxes since 1971, and battling the state of South Dakota over whether as a tribe member on a reservation, he had to have a license plate or driver's license.

Paul got off to a strong fundraising start. By mid-June 1987, before even getting the nomination, he'd pulled in around $150,000 from three thousand or so contributors. Paul's own mailing list was already at around 120,000 names by then, and his failed Senate run had raised $2.5 million in 1984. As libertarian magazine *Liberty* noted in 1987, if Paul could get even 2 percent of his list to jump ship from the Republican Party to the LP, the latter's membership would increase by 46 percent. Lonnie Brantley, a Texas libertarian who did phone fundraising for Paul in that campaign remembers it being particularly easy to do for Paul. "His house list has an unusually high compliance rate," Brantley said. "Seventy-five percent of what was pledged actually came in." A Students for Ron Paul group had active contacts in a dozen states. (As a sophomore at the University of Florida, I was part of such a group, and we brought Paul in to speak in January 1988.)

Paul for the first time was faced with an audience to whom

he had to prove himself sufficiently libertarian, rather than be defensive about being overly libertarian. He admitted that bills in Congress were often so convoluted and tried to do so many things that a clear proper libertarian vote wasn't instantly apparent to him, and perhaps occasionally in the 1970s and early '80s he'd slipped. He admitted that he had been more prone to vote for military spending during his 1970s stint in Congress than he should have been.

"The longer I was there the more government I voted against, including the military," he told *American Libertarian*. In some of his defense appropriation votes, he would think, "Well, I believe in defense and maybe I felt like it was the best to vote for that bill rather than vote for none. Later on I decided that it would be better to vote for none rather than a bill that I did not agree with."

The left-libertarians had lots of doubts about Paul, seeing him as a carpetbagger from the populist right. In the early days of AIDS, Paul was rumored to actually favor some sort of forced quarantine of victims; Paul denied that vehemently. Despite his long record, both before and after his LP run, of believing that border security and defense were fully justified on constitutional and libertarian grounds, he told the LP that while he may have some political issues with how open borders would play to the electorate, he did not object to the idea philosophically.

Paul's long-standing concern with monetary policy, hard money, and overspending and overtaxing led some in the LP to accuse him of only caring about those economic issues. He assured his LP audience that freedom to him was about more than just money: "The ultimate goal is to have a society where you don't have to worry about somebody knocking on your door and hauling you off." The idea that Paul was crazily obsessed with conspiracy theorizing already haunted him, and at an LP event he was asked to clarify his views about the Council on Foreign Relations (CFR), long a bugaboo of right-populists such as the

John Birch Society, fingered as the public place where the powers-that-be conspire against the liberty and property of the masses. While granting that he thought CFR folk are important players in national and international affairs, and that their philosophies undoubtedly *influence* world affairs, he denied believing that they "ran the world" in any movie-villain conspiratorial sense.

The LP had never seen such a tense, expensive fight over its hand. With Paul having raised a quarter of a million dollars just seeking the nomination, the politician ended up winning the bare majority needed for victory over the Indian rebel at the LP's convention by just three votes. The wisdom in the LP world was that Paul had won their minds—just barely—but not their hearts. Both his opponent Means and the man the party chose as Paul's running mate, former Alaska state congressman Andre Marrou, got more warm applause from the convention attendees than did Paul himself. Ron Paul's first presidential run had begun.

Candidates in third parties face problems that their major party opponents don't—exactly how the major parties designed it. Barrier one is ballot access. Every state has its own set of requirements for a political party to actually appear on the ballot, and generally it involves collecting many thousands of signatures, under complicated rules that vary wildly. The whole convoluted process costs lots of money and volunteer time. The Paul campaign became entwined with the LP's ballot access measures, and most of their money went toward that effort. In the end, Paul managed to appear on forty-six state ballots. In the aftermath of his LP experience, when Paul returned to Congress as a Republican, he retained his empathy for the plight of third parties, and has introduced a bill that would stabilize standards for getting on federal congressional ballots to a mere one thousand petition signatures.

Ron Paul then presaged Ron Paul now in interesting ways. Paul's fans were already flooding media with demands for Paul coverage—Phil Donahue and William Buckley complained about this—and early glimmerings of Ron Paul the man who can unify right and left were seen. A publication of the Socialist Labor Party defended their turf from a potential Paulist incursion, noting that his "simplistic stance can be appealing in regard to civil rights and foreign policy issues" but warning their comrades to evade his wiles. Paulites were already hanging banners in places they weren't wanted. At a baseball all-star game in Cincinnati, fans unfurled a banner reading: "Hey ABC! Why not Ron Paul for Baseball Commissioner?" George Bush was at the game, and someone whom the banner hanger insists she saw with his entourage ordered it taken down. (The hanger also insisted she got permission from the Reds before hanging the banner.)

Paul's campaign did not shy away from issues anathema to the right, such as withdrawal from both the drug war and the Persian Gulf. He had enough counterculture cred that the notorious 1960s LSD guru Timothy Leary hosted a fundraiser for Paul in his Los Angeles home. In one amazingly contentious and ferocious appearance with shock-TV host Morton Downey Jr. (a right-wing proto–Jerry Springer), Paul was a firebrand to an extent you just don't see from him anymore, pointing and shouting mostly in defense of drug legalization and peace. But he also relied on his existing allies on the conservative right who weren't satisfied with George Bush. Pat Robertson was 1988's Republican surprise, the place where flowed the always-present energy in the GOP grass roots, which found every non-Reagan candidate wanting to some degree. (That's a position that Paul finds himself in now, for different reasons. Pat Buchanan held it in 1992 and 1996.)

Robertson and the delegates who wanted him did not want George Bush. Paul's campaign pushed for a gaggle of dissatisfied Michigan delegates to switch their votes from Robertson to Ron Paul himself as a protest, but Robertson asked his delegates to succumb to party discipline and unity, hold their noses, and vote for Bush. One delegate was scheming to escort Paul onto the floor of the convention himself; Paul, disliking that sort of acting out, declined.

Paul saw in Robertson's people what Paul's people are now to the Republican Party: a new force, beholden more to their candidate and what they see him standing for than to the party itself. Paul could appeal to the religious right not just on the economic libertarianism and hard-money stuff—which resonated well with them then and now—but on social liberty issues such as free speech and just being left alone by the government to shape your own life in your own way. He could remind these people who valued homeschooling and the health of their own small religious communities that they should fear a government that interferes in their personal cultural choices—even if it means having to let the government respect choices they don't personally like.

The campaign ended in scandal: Paul's longtime aide Nadia Hayes was fired in the last week of the campaign and accused of siphoning more than one hundred thousand dollars out of Ron Paul's funds (from Paul's business accounts, not the campaign per se). Hayes ended up in jail for embezzlement.

Paul's vote total of 420,000 disappointed both his campaign and the party at large. As an LP analyst in the movement magazine *Liberty* wrote, their expectations were shattered because their expectations were unrealistic. Paul and Means both felt obligated to hype their possibilities in the quest to win the nomination, leading to ridiculous promises of 5–10 million votes.

Paul's campaign had promised national TV ads; they didn't deliver. One ad was produced and offered to local LP groups to air on their own local recognizance. It merely stressed Paul's status as the taxpayer's friend. An analyst in *Liberty* ran the numbers and found that where TV ads ran, they paid off: "LP vote totals in the counties where TV advertising was used ran 82% ahead of the pace in counties where no television was used." Paul said that the $3 million his campaign raised was only enough to cover staff, travel, and direct mail. (This desire for a Ron Paul presidential campaign to spend more on TV ads continued in 2008, coming from his frustrated fans.)

Paul himself was not daunted. In an interview with *Liberty* providing his personal postmortem on the campaign, the spirit that encouraged him to try again two decades later—and then try again despite a second failure—shines.

"If anybody should be disappointed or discouraged it should be me," Paul said. "I'm the one who was on the road for 18 months and I feel good about the whole thing. The people that called me, the big donors, thought the vote was almost inconsequential. They wanted to know what they could do, how to help the next project . . . so I just ignore people who want to be negative.

"I was making some calls the other day," he continued, "to people who had donated a good bit of money to me, and especially to one guy who had donated $5,000 to the ballot access plus the campaign . . . the guy voted for Bush. But he sent $5,000. In other words there is tremendous support, and sympathy and wishing us well, but this guy perceived Bush as being more libertarian than Dukakis.

"Now, I sometimes think Republicans are less libertarian than Democrats," said the once and future Republican congressman, "and they get away with more, but in his speeches Bush came out for a lot less government than Dukakis. . . . So I think that's a lot of what's happening out there: there's a lot of sympathy and wishing us well, but it has just not translated into a lot of votes."

And it still hasn't, for libertarians running with the Libertarian Party. This limns one of the tensions within the Paul movement still prominent today, between an effort to win political office and an effort to shift the consciousness of the American people toward liberty, regardless of the electoral outcome. While it is a mistake to say, as some do, that Paul neither expects to nor even *wants* to win the presidency, it is true that as he sees it, his campaigns can be a tremendous success whether he does or not. That's how he saw his 1988 LP run, and it's also how he sees his contemporary presidential runs—although that sense of optimism and looking toward the future of American liberty hasn't seeped into all of his supporters.

Paul was explicit that his politicking was about more than votes: "It comes down to the votes being irrelevant—although I don't like the totals, I'd like to have 2 million or something—but they're irrelevant as far as whether or not we're having an impact." Paul also announced that, in the quest for that impact, he questioned the libertarian purity attitude he embraced in his campaign, of refusing to seek out or accept federal matching funds to help pay for his campaign—although Paul sticks to that no-matching-funds stance today as a Republican candidate.

The Ron Paul on the other end of the 1988 presidential run was a more radical Ron Paul, and a more national Ron Paul. Those qualities propelled him on his next political moves.

———— ★ ————

CHAPTER FIVE

BACK TO CONGRESS

In 1995, Ron Paul decided Congress might need him again. He hadn't left politics or hard money after his failed Libertarian Party run. (He did more or less leave any active involvement with the LP.) Under the banner of libertarian financier and coin dealer Burt Blumert, Paul lent his name to a Ron Paul Coin Company, selling such items as "Ron Paul Survival Kits"—old World War II ammo cans filled with silver and/or gold coins. His newsletter business, with *Freedom Report*, *Investment Letter*, and *Survival Report*, continued. Under the tutelage of a new fundraising partner he met during the 1988 LP campaign, a man from the John Birch Society world named David James—sent Paul's way by old friend Jesse Helms—Paul in the early 1990s launched a FREE offshoot called the National Endowment for Liberty, which produced a TV news/commentary show hosted by Paul called *At Issue*, which aired on some cable channels. Paul was trying to emulate Milton Friedman's success with his *Free to Choose* series, explaining elements of his free-market message, one per episode. There were episodes dedicated to gun control, education, the drug war, the Federal Reserve, and noninterventionist foreign policy, among others.

In the wake of the end of the Cold War and the perceived failure of George H. W. Bush's Persian Gulf War, Blumert and other Paul fans decided Paul should try to run for president again, this time as the noninterventionist voice in a Republican Party that seemed genuinely torn between those who thought the Cold War's end meant America could come home and those with the neocon, American triumphalist view that it was time to fill the power vacuum left by the Soviet collapse and establish an American hegemony in the name of democracy. Pat Buchanan stepped up to take that role, and Paul, not thrilled with the idea of running again to begin with, backed out. Years later, Paul was comically critical of his own decision to even make the 1988 LP run. "Burt Blumert is a good friend," Paul told me in 2006, "but I'm still upset with him because he badgered me until I ran as a Libertarian in 1988."

Paul's old district, the 22nd, now belonged to House Majority Whip Tom DeLay. Paul had a beach house in Surfside Beach and used that as his official address and ran in the 14th District. It was a monstrous gerrymander, the 14th, then stretching over twenty-two rural counties from the Gulf of Mexico to the north and west, skirting around the big cities of Houston, San Antonio, and Austin. Democrat Greg Laughlin held the seat.

Well, *then*-Democrat Greg Laughlin. After Paul set up some outreach meetings with other Texas Republicans in 1995 about his intention to win that seat back for the party, Speaker of the House Newt Gingrich made a deal with Laughlin, promising him a seat on the powerful (and generally ultimately lucrative) Ways and Means Committee. Now Paul wasn't just running against an incumbent Democrat: he was an unwanted primary challenger, against an incumbent. Another challenger, Jim Deats, the last Republican to run against and lose to Laughlin in 1994, leapt in, too.

The Republican Party did *not* want Ron Paul to be a congressman again. The GOP establishment had a pretty good prima facie case, from Paul's own statements and behavior, that he was not one of them. As Paul remembered in a 2001 interview with *Texas Monthly*, "My image was completely different in 1996 than in 1976. You can't just get passed off as an average Republican having done what I did. We got hit hard."

In a 1999 interview with me, Paul talked about what possessed him to return to Congress. "The '94 results made me think there was a shift, a change in the air." He also liked the idea of the "challenge. I'd been labeled Libertarian, everyone knew my position on the federal drug war, civil liberties, knew how I'd vote. I wondered, could you still get elected when everyone knows exactly what you believed in? In the '70s they said they didn't know what I believed in, so this seemed like an interesting challenge."

Being hit by the establishment Republicans wasn't so painful for Paul. He thinks such outside interference helped him raise his $2 million, unusually high for a House race, in '96. "Every time someone from Washington came down here, I'd send out another fundraising letter and get another $100,000."

Laughlin got more than $1 million funneled to him from GOP leadership on the state and national level. Gingrich came to Texas to campaign for him, as did both George Bushes and Ed Meese, who tried to make sure everyone knew that Ron Paul didn't love Ronald Reagan enough. Paul had his friend and Texas icon baseball Hall of Famer Nolan Ryan as honorary campaign chair; he had his grandkids' precinct walking for him; he had an existing record of Republican attacks on Laughlin-the-Democrat before he became Laughlin-the-Republican-anointed to remind voters of.

That didn't win the primary for Paul, at first. Laughlin smacked Paul down 42 percent to 32 percent, with Deats at 24 percent. Alas for Laughlin, he needed a clear majority for a primary victory. A very low-turnout runoff between just Laughlin and Paul followed—with Deats openly endorsing Paul—and Paul won with 54 percent.

Paul's opponents, both in the primary and then the general election against Democratic lawyer and union favorite Charles "Lefty" Morris, painted him as a madly radical libertarian who would be for selling heroin in public schools, if only he believed in public schools. "People would think, 'a small-town baby doctor could not be everything Laughlin painted him as' and throw the [anti-Paul] mail in the trash," said Paul's old congressional aide Eric Dondero in a 1999 interview. (Dondero turned very publicly against Paul later over Paul's noninterventionist stance after 9/11, and despite being quick to critique Paul in public forums on the Internet, he declined to be interviewed for this book.) Mike Holmes recalls ads against Paul trying to portray him as a loon—"weird noises, booooiiing!!!, cartoon-like effects, he must be *crazy*!" But anyone who actually sees the quiet, calm, wonkish country doctor knows that's nonsense. Any personal contact, Holmes says, disabuses a voter of the "idea that he's any kind of dangerous nut. They tried to paint him as a dangerous extremist, but meet him and he's painfully ordinary." Holmes remembers fondly a press conference during that 1996 race in which, of course, Paul was hit with the drug question. "Ron said: repeal the Harrison Narcotics Act. That's always been his position and he didn't waffle it or attempt to change it. I thought, Ron Paul's back, and he has not changed. That's why he has such a loyal following; people have always been willing to walk on water for him."

Paul outraised Morris handily, around $1.2 million to Morris's $472,000. Through his entire congressional career, thanks to his national fan base, Paul has almost always raised twice or more than the national average for congressional races.

The fight against Morris featured the first appearance of the greatest scandal of Paul's career—the only one with any teeth for a politician with a reputation for consistency and integrity. For a few years, the people ghostwriting Paul's *Survival Report* newsletter indulged in some occasional mean and contemptuously jokey references to blacks. Among the quotes circulated to Texas papers, from a 1992 edition of the newsletter, were "I think we can safely assume that 95 percent of the black males in [D.C.] are semi-criminal or entirely criminal" and "Opinion polls consistently show that only about 5 percent of blacks have sensible political opinions."

Various Texas papers picked up on the newsletter controversy, and at the time Paul merely wrote the comments off as both "tongue-in-cheek" and "academic" (which do not exactly go together, and while the first one clearly applies—this sort of comment was the bread and butter of resentful right-wing radio of the era—the latter really doesn't), he did not disavow them or deny them per se. (When the scandal resurfaced in 2001, he did, which will be discussed in chapter 6.) The whole brouhaha didn't cripple Paul electorally; he beat Morris by a tight 51–49 margin, so tight that all through the election night Morris was denying he'd lost.

Paul went on C-SPAN shortly after his new term began and in response to a question about the 1993 Waco incident, where federal agents searching for illegal weapons laid siege to and eventually killed more than seventy members of an eccentric religious community called the Branch Davidians, the new congressman from Texas said "there's a lot of people in the country who fear that they may be bombed by the federal government at another Waco." Democratic representative Chet Edwards, in whose district the attack occurred, called Paul's comments "sheer lunacy at best." Ron Paul was back, and Congress wasn't going to forget it.

For Paul, dedication to the public purse starts at home. A congressman again, he's never voted for congressional pay raises. He refuses to participate in Congress's very generous pension plan, and he regularly saves enough on his allocated office budget to return money to the Treasury.

Paul was not given any seniority credit for his prior term, though you typically would if the leadership wasn't mad at you, and he did not get his choice of committee picks. He ended up on Education. Paul thinks Gingrich was loath to have someone with Paul's foreign policy positions on a foreign affairs committee under his watch—Paul's opposition to all foreign aid in particular he understood had gotten him blackballed. In 1999, an Education Committee staffer told me that a lot of the "Education and Workforce people hate him, because his attitude is no, no, never, no." It was "largely moderates on that committee, wondering, 'how can we leverage federal dollars to make education better?' And Ron's like, 'what do we need public schools for?'"

By the time the second Bush era began, Education was where the White House was least happy to have an outspoken foe of federal involvement in education. That, one Paul staffer thinks, had a lot to do with Paul finally getting on Foreign Affairs. "Using [Education] as a platform to publicize libertarian opposition to the president's signature domestic initiative" would have been less than ideal, and "by giving [Paul] what he really wanted, it made sure he wouldn't complain."

Paul did get back on Banking, and two of its subcommittees: Domestic and International Monetary Policy (the perch from which he could snipe at Federal Reserve chiefs) and Financial Institutions and Consumer Credit. Even in 1999 Paul told me, "I'm getting prepared because there will be a bubble burst here. The Fed won't be held in quite so high regard. Last time I interviewed [then-Federal Reserve Chair Alan] Greenspan before the committee, everyone was praising him on high, practically bowing. Why couldn't they reappoint him immediately? I flat out told him I see this bubble coming, I don't understand why you don't get out while the getting's good. He just laughed. He knows what I think I know. He's studied all this."

Paul became notorious for his (now prescient) hectoring of Greenspan over the next decade; Paul was especially aggravated because Greenspan was an apostate from gold, having written during his Ayn Rand acolyte days in the 1960s an essay about why a gold standard was best. "I dug out my copy of *The Objectivist Newsletter* [edited by Ayn Rand]," Paul says, "where he wrote his gold article [in which Greenspan praised the gold standard as a source of economic stability, guarantor of economic liberty, protector of savings, and check on government's power to inflate and spend]. When we started talking I flashed it out and said, 'Remember this?' He said, 'Yes, I certainly do.' I opened it up to his article and said, 'Remember writing this article? Would you autograph it for me?' And while he was autographing it I said, 'You want to write a disclaimer on it?' 'No, I read it recently,' he said, 'and I wouldn't change a word.'

"Toward the end of his reign I brought that up again in a congressional hearing. I was a little more confrontational with him about what he used to think and why it's different now, and he said, 'That's a long time ago, and I no longer subscribe to those views.' He *did* put a disclaimer on it. The first encounter was private, and the second was a public statement."

I first wrote about Ron Paul early in his return to Congress, in a 1999 article for the conservative magazine *American Spectator*. It still sums up a lot about his relationship to his party and to Congress:

In Congress party discipline is vital. Leadership craves a tight ship where all hands can be counted on to row in the same direction. So many congressmen bridled when, after stressing the need for everyone to pull together to get a conference bill through, then-Speaker Newt Gingrich invoked a "Ron Paul exemption," granting the gentleman from Texas's sprawling 14th District a hassle-free bye on the vote.

This story is part of the growing legend of Ron Paul, the Exceptional Republican. Though his name rarely appears in the national press, and his face almost never on Sunday morning news shows, in 1996 he was third only to Gingrich and Bob Dornan in individual contributions to Republican House candidates. While he hasn't managed to get any of his own bills out of committee since re-entering the House in January 1997, he's considered a vital asset by a large national constituency of libertarians, goldbugs, and constitutionalists. He's defied one of the holy shibboleths of electoral politics— Thou Must Bring Home the Bacon—by being a consistent opponent of agricultural subsidies in a largely agricultural district, and he's still won twice in a row.

[After the bruising fight with Laughlin and the national GOP,] both Paul and GOP leaders deny any animus was involved, though Paul still makes asides about "establishment Republicans (who) want to dance on my political grave" in his fundraising letters. The official explanation is that the GOP craved Democratic defections, and needed to prove the party would go to bat for them. At the time, though, Laughlin insisted that "Republicans who have worked with Paul are appalled he is in this race. When they hear about his wacko ideas, they get even more concerned."

Paul insists there are no hard feelings, and it's hard to find GOP colleagues who will openly say bad things about him as person or legislator. . . . Jim Leach, chairman of the Banking Committee, says Paul is "perhaps the most philosophically thoughtful and personally decent member of Congress. We don't vote the same way, but he's the nicest guy around." Bill Goodling, the Education and Workforce chairman, says—

and one can almost see the bemused smile between the lines of his written statement—that Paul "has brought a different perspective to the Committee . . . (H)e sticks to his own moral compass, and political considerations do not influence his decisions on how to vote. . . . He is very predictable: If proposed legislation expands government or involves activities which he does not consider specifically authorized by the Constitution, then he will vote No." . . .

Indeed, as libertarian-leaning GOP coalition builder . . . Grover Norquist of Americans for Tax Reform says, one Ron Paul is grand; and 218 Ron Pauls would be even grander; but 20 Ron Pauls could cripple the party, since the usual half-steps toward less government and less taxation might not find support among the more ideologically rigorous.

"Some Republicans don't work with the rest of the gang because they are being jerks, or playing to the home team, or being weak," Norquist says. "Ron is understood to be acting on principle. But he does take principled positions that sometimes cause the leadership heartache because they need to pass less-bad bills, and they can't count on his vote to do that."

It's a bit of a legend spread via Gary North and Paul himself that he refuses to hire people with previous Capitol Hill experience—when he first got there he did hire the Koopmans, from Steve Symms's office—but he does try to avoid it. There's a mentality endemic to the would-be lifelong Hill rat that isn't very useful around Ron Paul.

"People who have Hill experience can be accustomed to certain things that don't fit with how Ron likes to operate," his long-time legislative director Norman Singleton says. "Sometimes they are just looking for the next job, so they're afraid of offending someone on the Hill, particularly leadership. Hill professionals tend to think their job is to keep their boss out of trouble by avoiding controversial votes or introducing controversial legislation." Thus, when Paul got back in January 1997, his crew was inexperienced.

J. Bradley Jansen, who worked as a legislative aide on banking issues for Paul from 1997 to 2001, recalls that only Paul's scheduler had any Hill experience whatsoever. "No one in the office knew anything, and when I say anything, I mean *anything*," Jansen says. "I had an appointment for a Banking Committee meeting in Rayburn 2128, RHOB, and I didn't know what RHOB stood for [Rayburn House Office Building]. None of us knew anything about the Hill. And unlike the other offices, we didn't just follow whatever party leadership was saying, and most staffers rely on the party for a lot of background on issues. We had to do everything ourselves, instead of copying Republican National Committee talking points."

Jansen helped Paul with one of his early high-profile successes as a congressman: the fight against so-called Know Your Customer banking regulations, which would have required banks to form a "normal" profile of your financial behavior and report any irregularities. "This was an old and honorable term in banking, 'know your customer,'" Jansen says. "It's an antifraud thing, meaning someone can't come in and say 'I'm Brian Doherty, and I want to withdraw all my money.' They tried to redefine that for this huge bank spying program, requiring them to determine the source of funds from whatever money you deposited."

Paul's office organized a broad-based coalition against the regulations, including antitax and antiregulation groups from the libertarian-leaning right. "It was done in the name of the war on drugs, to fight money laundering supposedly, so I went to the drug policy groups. I got banking associations on board, a lot of the Christian right on board because they didn't like Clinton," Jansen says.

Jansen recalls an "only in Ron Paul's office" meeting on

Capitol Hill, with a representative from Concerned Women for America sitting next to a representative of the Marijuana Policy Project sitting next to a representative of the Independent Bankers Association sitting next to "some lefty consumer privacy group" sitting next to the Competitive Enterprise Institute. Paul's office managed to generate hundreds of thousands of public comments opposed to the program, Jansen says. While that law was defeated, in the wake of 9/11 most of its negative provisions have been enacted under other cover—a point Paul indeed often made after the attacks about the rush to control.

Paul's bills, famously, don't tend to go anywhere, but a brief survey of some of them helps show where Paul is coming from. He's introduced legislation to abolish the Occupational Safety and Health Administration, repeal the Gun Control Act of 1968, end the postal monopoly, restrict the IRS's ability to inspect our tax returns, repeal the National Flood Insurance Act, make paper dollars legal tender *only* for taxes, repeal Selective Service, impose term limits on legislators, audit the Fed (many times), withdraw all the military from Europe and Japan, end the Fed, prohibit the use of federal money to enforce United Nations restrictions on Iraq (in 1997!), or in Kosovo (in 1999), bar federal teacher certification, legalize hemp, and give tax breaks for stem-cell research (unusual for a pro-lifer).

The number of ways a hard-core libertarian might want to change the laws of the United States approaches infinity, but Paul tends to concentrate on a few areas. Most of them deal with his central concerns of monetary policy, foreign policy, and civil liberties. He also has more charming little ideas that his fans enjoy, like the Tax Free Tip Act, which would exempt servers' gratuities from income taxation. Paul fans make little business cards about this idea and lay them out along with their tips on pub crawls, spreading the word of Ron Paul to bars and restaurants across the nation.

Paul is fond of that sort of targeted tax break or exemption, and he's offered them for teachers, firefighters, cops, and money spent on tax attorneys, among others. This sort of thing drives some free-market economists crazy. They worry about the distortions and misdirections in economic decision making that come from allowing certain economic activities to be cheaper than others because you aren't taxing them. This doesn't worry Paul. He tends to take the blunt, frontal approach on taxes: that it's always good to reduce them, anywhere and everywhere you can, and allow Americans to keep more of their money and decide where and how it will be spent. "From our perspective," says Paul's congressional chief of staff, Jeff Deist, "any libertarian should be all for cutting any tax on anyone for any reason any time. Does it reduce or eliminate an existing tax for someone, that's the criteria. Does the tax code reward and punish? Yes, it does, unfortunately," but Deist believes that should not override the imperative to reduce the amount the government takes out of the economy.

"We're not naïve," Singleton insists. "We understand that no Speaker of the House has been calling us to say, hey, what infringement on liberty have you guys heard about that needs to be addressed legislatively? And if you give us some Ron Paul bills we will put them on the floor next week. But we are very proactive and we are trying to use legislation not just to pass this Congress, but to educate and make a point."

Sometimes Paul's concerns have intersected wider ones, and often presciently: years before the Real ID Act passed Congress (it has since been more or less scuttled by states' refusal to go along), Paul in 2000 introduced a Freedom and Privacy Restoration Act to prevent any sort of national identification card from being required for such things as buying a gun, opening a bank account, or traveling. The post-9/11 era has swiftly and thoughtlessly overturned many ingrained cultural suppositions. "Your papers, please" was considered such a staple of totalitarianism it had become a jokey signifier that everyone recognized. The lessons of the boy who cried wolf were turned on their heads by those who insist, in the face of no evidence, that we face a persistent and horrific threat from domestic terrorism that demands a high degree of restriction on our freedom to move and our Fourth Amendment rights. Now they are the sober, sane ones and people who point out we haven't seen any sign of this so-called wolf for over a decade are seen as the irresponsible loons. Paul's concerns for personal privacy and liberty and a humble foreign policy have all been on the losing end of these cultural shifts. But the very fact that they seem to be losing energizes those who share Paul's concerns to support him so fervently.

Ron Paul has always been Ron Paul. Take a look at his record after his return to Congress. An end to international entanglements, check—he was once again trying to get us out of the UN. An end to an imperial presidency, check: in July 1997 Paul was griping about congressional acquiescence to Clinton's military adventures in the disintegrating Yugoslavia. Already in January 1998 Paul was pointing out the unconstitutional folly of our Iraq policies, telling the House that, even though it might be easy to score cheap points against Clinton by pointing out that his planned attack on Iraq seemed nicely timed to distract from the Monica Lewinsky scandal, the real problem was "a flawed foreign policy. . . . [T]here was a time in our history that bombing foreign countries was considered an act of war, done only with the declaration by this Congress. Today, tragically, it is done at the whim of presidents and at the urging of congressional leaders without a vote." That same month he noted on the House floor that

> the chaos that we contribute to in the Middle East assures me that there is no smooth sailing for the new world order. . . . If significant violence breaks out, it will cost American citizens money, freedom and lives. . . . I see our cities at a much greater

risk than if we were neutral and friends with all factions. . . .
the way we usually get dragged into a shooting war is by some
unpredictable incident, where innocent Americans are killed
after our government placed them in harm's way and the enemy was provoked. Then the argument is made that once hostilities break out, debating the policy that created the mess is
off limits. Everyone must then agree to support the troops.

Paul was still very much a man of the populist right in the late
1990s, very worried about sovereignty, globalist efforts through
the UN, other international organizations, and unrestricted border crossings to destroy our uniquely American identity and institutions that we had direct control over. The Paul now running
for president stresses these issues not at all; our current internal
problems of debt and imperial overreach seem much more important a problem than certain decisions being made by bureaucrats
outside of Washington. In a March 1998 lecture to Congress,
Paul sounded out our foreign policy future: "Mr. Speaker, last
week it was Saddam Hussein and the Iraqis. This week's Hitler is
Slobodan Milosevic and the Serbs. Next week, who knows? Kim
Jong Il and the North Koreans? Next year, who will it be? The
Ayatollah and the Iranians? Every week we must find a foreign
infidel to slay, and of course, keep the military industrial complex humming." Jackie Gloor, who runs Paul's Victoria, Texas,
office and has been with him since his return to Congress, says:
"Who would've thought people in sleepy little towns in south
Texas would be talking about things like the military-industrial
complex? But they do because all of a sudden we understand it"—
because their congressman explains it to them.

Paul attacked Clinton's militarism in both Yugoslavia and
Iraq, condemning the Iraq Liberation Act, which set the stage for
further U.S. military intervention there in the name of supporting democracy. It did not all start on 9/11, as Paul remembers all

too well. He was complaining all the way back in 1998 about our involvement in giving money, technology, and training to Osama bin Laden.

Paul voted for Clinton's impeachment but used the moment as an opportunity to tweak his colleagues: "There is a major irony in this impeachment proceeding. A lot has been said the last two months by members of the Judiciary Committee on both sides of the aisle regarding the Constitution and how it must be upheld. But if we are witnessing . . . a serious move toward obeying the constitutional restraints, I will anxiously look forward to the next session when 80 percent of our routine legislation will be voted down."

How did a guy like this keep getting reelected from small-town, rural Texas? I asked Jackie Gloor, native Texas woman, with that great Texas kind toughness, who worked for Paul's primary opponent Jim Deats in 1996. After the runoff came down to Paul and Laughlin, Deats told her that they both had to work for Paul's victory. She began precinct-walking for him and has been on his team ever since. She has many deeply felt stories of his personal kindness and decency as boss and friend, and believes "everyone loves Ron Paul" when they actually meet the man. She knows many people hate his *views*—enough of them are calling her office all the time to complain. And for sure, his district isn't made up of hard-core libertarians.

"He educated this district from the very beginning," Gloor says. "I think the first tough vote he had in '97 was when he voted against making it a federal law to burn the flag, or to desecrate the flag, and the people were in an uproar about it. And he didn't run from it. He came right down here, he went to every little town and explained why and stood up in front of people and was ready to take their criticism, and they were there to bring it! And he was there to answer it and this district suddenly began to understand how important it was that we truly have freedom of speech and expression. And that it's really a property rights issue. If you have a flag you bought that flag, or it was given to you and it belongs to you, you have a right to do with it as you will. Even if it makes someone else mad. They have a right to be mad but you have a right to do with it what you want to with your own property. And before Ron, that never would have happened.

"And he has done that time and time and time again. I cannot tell you how many times he's taken a tough vote and then he comes right back to the district and explains it to where we understand. We may not agree with it, but we learned even if we didn't understand a vote we could trust his heart and we knew that what he was doing he truly believed in it. If he voted for it, it passed constitutional muster, if he didn't vote for it, it didn't. And that was just pretty much it."

Gloor has personally felt the transforming hand of Ron Paul. "I came in a flag-waving, country-loving American citizen that thought my government did nothing wrong. Boy, has that changed. I see everyday abuses that this government does to people—from people being convicted of a federal crime for watching pornography on their computers at home in the privacy of their own rooms, and they go to prison for that. The war on drugs has totally changed my opinion—I have never even smoked a joint, but I totally believe that drugs ought to be legal and that people ought to be allowed to make stupid mistakes if they're going to suffer their own consequences. And I'm one that thought people that bought marijuana on the street ought to go to jail. I don't feel that way anymore, and it's because I've seen firsthand now the abuses of our system. If you have money and you're white and you live in south Texas you get off, if you can afford a good attorney. And if you don't, you go to jail. It's wrong. It's just flat wrong. And this job has taught me that, I wish it hadn't, I think naïvety has got a little bit of an *ease* to it. Our government abuses people all the time and now the president's taking out hits on U.S. citizens abroad and people think that's okay. That's scary."

"Ron violates so many of the rules people have on the Hill," says Singleton. "Like, never vote by yourself, or take a position against the majority of your party. Never vote against your district's interests. We used to have people asking us, 'What's your district like?' I think they expected the answer to be, well, imagine Galt's Gulch [the home of Ayn Rand's libertarian heroes in her novel *Atlas Shrugged*] except a little more Christian."

Jansen remembers envy from the staff of other right-leaning congresspersons. "They'd be frustrated with their own boss, and in their minds Dr. Paul was always voting the right way, and they wanted their boss to vote the right way. They thought their bosses knew better but just gave in to important constituent pressure or important donor pressure or pressure from leadership, fear that Newt would come down hard, and didn't vote the way they knew their congressman should."

The standard explanation for Paul's political success: very efficient constituent services of the "help me deal with Social Security, immigration and IRS problems" variety, and always the four thousand babies he birthed. Gloor says that kind of thinking discounts Paul's political prowess. Most of the people who come in with their problems are likely not even voters, she thinks; it isn't *them* sending him back to D.C. all the time, by higher and higher margins. They just know that when you've got some problem with the government, you come to Dr. Paul.

Local Republican officials are ambiguous about Paul. None of them want to complain about him on the record; one who through a long conversation was dismissive of him as naïve and overly obsessed then revealed in the end he's never voted for anyone else. Still, he finds Paul's socially laissez-faire attitudes naïve. When he sees Paul object to the government pressure on private business inherent in the 1964 Civil Rights Act, and maintain that the forces of social pressure would insure that hardly anyone would discriminate anyway, "I'm thinking, what truck did he fall off of? Of course people will do that. I don't think he's any kind of a bigot or racist or anything like that. I just don't think he's been around the kind of people I have been around. Not the type who went to medical school, you know?"

Another local observer notes that locals are furious about Paul's refusal to help the county get more aid after Hurricane Ike hit in 2008. Another local Republican official notes that a mechanical monkey set up to hit a "no" button could be as effective for his district as Paul.

Paul, in a 1999 interview with me, seemed to treat his political success as a mystery himself. "Because of the nature of my district and my views, people tried to assure me my views did not get elected. They don't know exactly how you feel and if they did they wouldn't elect you. So I stick to my guns and let them know what I believe. Nobody is quite sure exactly why I win. . . . [W]hen I take a vote contrary to a prevailing attitude, instead of hoping no one will notice I send out a press release."

It doesn't hurt that Paul can always outspend his opponents because of his national fundraising base. That's not because of anything unique about his district. Paul has, Singleton points out, "a fundraising base independent of party machinery." And that's vital to allowing Paul to be Paul, never mind the question of the *will* to be Paul, which even most supposed superconservatives in the Republican Party lack.

"I remember in 2000 I went to a briefing on how conservatives in Congress should deal with a Republican president who would not always take the conservative side and ask us to vote against the way that we wanted to," Singleton says. "One of the things the guy directing the meeting said was that Karl Rove will start calling your boss's big donors and saying, 'Why are you voting against the president?'

"I told Ron and the guy who did most of Ron's fundraising at the time, David James, about this conversation. They both started to laugh. David said, 'I wish Karl Rove would have conversations about the president's policies with some of our donors. He might learn something.'"

That was 2000. By the end of 2001, everything had changed. Paul has warned us decades ago about the problems that could arise from putting our troops in Saudi Arabia and he had warned us years ago about the problems that might arise from our hostilities in Iraq, and yet no one seemed to be listening. The U.S. government's reaction was everything Ron Paul could have feared: the revival of old civil liberties restrictions (including the "know your customer" banking regulations he'd previously defeated), baying for and launching aggressive wars, and a refusal to consider simple liberty-respecting expedients such as arming pilots or using letters of marque and reprisal to hit our nonstate terrorist enemies.

He kept getting sent back to Congress, but Ron Paul was not winning.

———— ★ ————

CHAPTER SIX

THE RON PAUL REVOLUTION

As the Republican Party became a bigger-spending, more warlike, and more awful entity throughout the mid-2000s decade, many people started thinking about bugging Ron Paul to run for president. Some of the old Texas Republican Liberty Caucus crew got a meeting with Paul to pitch him on it in 2006, but he said no. Nevertheless, he let himself be persuaded eventually; he credits Kent Snyder, an old Paul hand who worked with the Liberty Committee, a legislative lobbying group Paul has since broken with. When I interviewed Paul about the campaign before it had even officially begun, Paul told me, "It will have to be a grassroots campaign and rely on the Internet. If we don't learn how to use that to its maximum benefit, we won't have a very viable campaign."

They learned. Justine Lam was an early hire. Lam had previously worked with the Koch-funded Institute for Humane Studies, as well as with the Koch Summer Fellow program, where Paul was frequently a speaker. She remembers being highly impressed when he answered a question about how best to be a libertarian in electoral politics by suggesting that anyone who actually *wanted* to be a politician probably was the wrong person to be a politician.

The operation didn't even have an office yet, and so they worked out of Snyder's home. Out of impecuniousness—doing things online didn't require expensive office space—Snyder firmly encouraged Lam to do what Paul already knew had to happen: look to the Web to bring the campaign to life.

That didn't mean whipping up some super Ron Paul 2008 site to draw in the world, Lam says. "We wanted to spread the word out. People wouldn't come to our site unless they already know, so we want them talking about Ron Paul on other sites, YouTube, MySpace, pictures on Flickr. We didn't want to encourage supporters to spend all their time on our site. By decentralizing it we got people to use other sites, and each other."

Paul developed an earned reputation as the man with nineteenth-century ideas (for those who still mocked his classical liberalism as hopelessly outdated) who was mastering twenty-first-century techniques for communicating and rallying the troops. The campaign's own weaknesses in conventional terms became strengths. With a campaign that didn't have enough money to do everything, Lam says, Paul's supporters were "forced to look at themselves for resources and not us. And we'd tell them, 'we can't do anything with your idea, but you are welcome to do anything you want. But don't tell us so we don't have to report it to the FEC.' But we encouraged the fans to use their expertise and creativity in spreading the message about the candidate.

"And they saw these new Internet tools were important to spreading the message. Many smart people understood the power of YouTube. Then people saw others doing videos, with these emotionally appealing videos' imagery, and they started doing it themselves. We saw a lot of iterations of people testing things out with no idea what was going to work."

The YouTube avenue was important. Very important. The

most common answer to the question "How did you get into Ron Paul?" from his devoted activists was, without question, "Someone sent me a YouTube video." They don't necessarily always remember which one specifically—by now most of them have seen dozens if not more.

A strange variety of viral videos has spread around YouTube from early 2007 until today, many featuring unofficial Ron Paul campaign songs. The range of styles in these Ron Paul ballads reflects the eclecticism of the Ron Paul Revolution: from wan old-school folk to 1990s-style jazzy trip-hop, from sprightly garage rock to straight Sinatra steals. Some lyrical samples, from the trip-hop number: "We need Ron Paul / For the long haul / 'Cause he'll stop all the wars / Where the bombs fall." From the garage pop tune: "Ron Paul! / He's got brains and he's got balls / Ron Paul! / Who you gonna cast your vote for next fall? / Ron Paul!"

There was the Ron Paul Girl for cheesecake and comedy, and then twelve different Ron Paul hotties for the fundraising wall calendar; there was the Granny Warrior crossing the country for Ron Paul with her pet monkey; and then the apotheosis, the video featuring a rapping pizza slice singing the praises of Ron Paul. I was never sure if that one wasn't a parody, but in the end it didn't matter.

Jonathan Bydlak, a former financial analyst from Connecticut, says he's a classic embodiment of one of the most popular homemade Ron Paul signs as the campaign took off: "Dr. Paul Cured My Apathy." Bydlak was an early hire running the fundraising operation for the campaign, which he recalls as at the very start just "Kent Snyder sitting in his apartment posting YouTube videos of Ron Paul speeches on the floor of the House.

"People look back and say [the fundraising] was all done independently of the campaign, and I will say, I probably deserve five to ten percent of the fundraising credit and not more," Bydlak says. "But a lot of things the campaign did enabled the good things to occur. I look at the meetup groups." The campaign made a deliberate decision not to try to reinvent any existing wheels when it came to online organizing. The Web tool Meetup.com had already created a means for people to communicate and organize themselves around an interest. The campaign decided to just encourage their activists to use it, and it worked. By the end of 2007, more than sixty thousand Paul fans were organized through Meetup, far more than any other candidate.

Paul's crushing dominance when it came to any measurable show of online support—meetup groups and members, YouTube views—became a key aspect of the occasional, confused, head-shaking media minding of Paul. Everyone knew that if they mentioned Ron Paul in a manner that any fan even *thought* disrespected him (and Paul fans are fiery of heart and have far more sensitive meters for disrespect than the average person), you could count on hundreds if not thousands of perturbed emails and posted comments. Jan Mickelson, a prominent and respected right-wing Christian radio host in important early caucus state Iowa, had Ron Paul on frequently and learned firsthand that Paul fans are "crawl-over-broken-glass zealots—fiercely devoted, they love him, they are passionate, wherever he appears they appear, wherever he's on TV they watch, whatever poll they can participate in, they respond. If you get on their right side, you are friends for life; if you nuance even a little bit your support for him they come at you."

Any online poll, where the sheer numbers of passionate fans sitting by a computer insure a win, would be won more than handily by Ron Paul. Then the campaign thought of using the meetups and their camaraderie as a fundraising tool, setting up national contests to see which local meetup could raise the most in a set amount of time, with a Ron Paul appearance as a prize. Rather than just being cutthroat, people in different cities' meetups began cross-contributing to each other. "If that meetup thing hadn't happened," Bydlak says, "I don't think anything else subsequent happens" in terms of their fundraising success.

Despite the slow spread of Paul ideas on the Net in the spring of 2007, the real explosion came with the Giuliani moment in May, as described in chapter 1. Paul made immediate hay out of it, throwing a news conference at the National Press Club in which he jabbed at Giuliani's ignorance about radical Islam, our history in the Middle East, and what Bin Laden and al-Qaeda had actually said about 9/11. It included the government's own *9/11 Commission Report*, and three books by scholars from the shadowy world of truly realist foreign policy that the staunchly anti-intervention Paul moved in: Michael Scheuer's originally anonymous *Imperial Hubris: Why the West Is Losing the War on Terror*, Chalmers Johnson's *Blowback*, and Robert Pape's *Dying to Win: The Strategic Logic of Suicide Terrorism*. These books, by people who'd worked or studied extensively in the field of our enemies' thoughts and actions, explained that history matters, that actions have consequences, and that the bloody terrors of radical Islam exist not in a vacuum but in a reaction to our own ill-considered interventions and alliances in the area.

None said that we deserved what we got or were to blame for the crimes of others, and neither did Paul. But they did say that in trying to solve a crime, motive is important, and that empire has costs and discontents that need to be intelligently considered. From understanding them, Paul felt justified in what he's always believed: that the United States would be better off minding our own business, as a domestic and mercantile republic, not a world-spanning martial empire.

Thanks to the new fans Paul attracted by going very public with his noninterventionism, the campaign came in with a surprising third-quarter take of $5 million, whereas the media had been anticipating no more than $3 million from this strange, obscure campaign. "Our online fundraising successes were important not just because of the money itself, but as a big news story in and of itself," Bydlak says. In the meetup-winning city, Philadelphia, more than five thousand people came out in early November to see a candidate that the media still mostly pretended did not exist. Paul joked about how he hears his supporters are just a handful of computer spammers—but it looks like there are a lot more!

The campaign was experimenting with website widgets that showed donations rising in real time, with names of donors displayed as people donated. This became a huge game and thrill for fans, waiting to screenshot their names, and helped lead to a million-dollar week. They'd display a quill pen being filled with each 1,776 donors, and fill three in a week. Snyder loved how transparency led to cash, and set a public goal of $12 million by end of the year. He wasn't afraid of the humiliation of setting an unreachable goal; why not go for broke? They learned, Lam says, that a $12 million graphic doesn't rise fast enough to excite people.

A couple of Paul fans, Jesse Elder and Eric Nordstrom, came up with the concept of the "moneybomb" (encouraging everyone to give on one day) and created a "ronpaulmoneybomb" website. They promoted one for an impressive October 20, 2007, take of around $145,000. Another Paul fan, James Sugra, was also thinking of the power of small numbers to add up to something big, and began imagining that if a *million* Paul fans could be mobilized to each give ten dollars on one day, well, that's a $10 million day. The website graphic would sure rise noticeably *then*.

Sugra made a YouTube video hyping the idea, then another activist, Trevor Lyman, stepped in to gin up a very successful cross-the-Paulverse promotional campaign for the idea. Paul fans loved the rebel message of the dystopian graphic novel and film *V for Vendetta*, about a revolt against a futuristic totalitarian Britain, and which uses iconography related to Guy Fawkes, the man who on November 5, 1605, attempted to blow up Parliament in what became known as the foiled Gunpowder Plot. The anniversary of that plot became the new moneybomb day. That was edgy, linking a presidential campaign to violent rebellion, even against a fictional English dystopia, or the British monarchy of four centuries ago. But hell—wasn't America *born* in violent rebellion against the British monarchy?

Sugra's $10 million fantasy never happened, but the Ron Paul world was still rocked by the $4.2 million that came in that one day. Sugra's original idea still seemed solid: why *not* try for $10 million? An energized grass roots began holding on to their coins, waiting to throw them on another day rife with rebellious significance: December 16, 2007, the 234th anniversary of the Boston Tea Party.

The timing wasn't great for the campaign. December 16 was a bit late for optimal Iowa and New Hampshire advertising action. But attempts to lean on the grass roots to move it forward failed—in a *Los Angeles Times* feature on Paul fanatics, campaign fundraiser Bydlak was condemned as an "idiot" for even trying— and the Tea Party moneybomb became the biggest single-day fundraising for a political candidate in American history at the time, with over $6 million pulled in. The explosion of these moneybombs propelled Paul to the top of the GOP field in fourth-quarter 2007 fundraising.

Some noted an irony in a movement of individuals taking such joy and finding such fulfillment and success in a collectivist movement, as part of something bigger. The difference—the key to the whole Ron Paul phenomenon—is that the moneybomb is voluntary, and as any Paul fan would tell you, the joys of freely chosen collectivism beat those of the forced kind.

Bydlak stresses that the advantage of the moneybombs went beyond the money, even beyond this very public demonstration

of the basic Ron Paul message of decentralized but highly effective spontaneous order. The real gold in the moneybombs was the
number of new donors it brought in as a raw number of people—a
raw number now invested, now on the campaign's lists for future
dings, now more likely to bug their office mates or families about
their candidate.

The weird *timing* aspect of the moneybombs got a little tricky
for the campaign, though. While he is sure "the benefits out-
weighed the costs" of the all-in-one-day strategy, Bydlak does say
that "in hindsight I might message more to make them see the
goal is to win the campaign, not necessarily raise the most money
in one day. There's a time value to having money sooner," especially with the world-record December 16 take.

The media thought Paul was just being silly during that first
phase of campaigning in 2007, as did most of his colleagues. For
a while at least. As Paul's congressional legislative director Norm
Singleton recalls, "he was lumped in with [Tom] Tancredo and
[Duncan] Hunter as a nonserious candidate who is basically running to make some points, but isn't going to get anywhere." But as
the revolution took off, "I'd have [other congresspeople's] staffers
talking to me about going home and seeing Ron Paul signs all
over their boss's district, and trying to not let me see how shocked
they were, but I can tell they were shocked.

"Then what really began interesting a lot of these people on
the Hill is fundraising, like when Ron started outraising McCain
and Giuliani. That really got their attention."

The strangest result of Paul's run, especially because of how
much it became linked to a foreign policy of peace thanks to
the Giuliani moment, was the positioning of Paul in many supporters' minds as being to the supposed "left" of all the Democrats, except perhaps Dennis Kucinich (for whom Paul always
expresses admiration). Singleton says a friend of his claims that
"Ron can take credit for saving a good chunk of politically aware

and active young people from socialism. By running as the most hard-core antiwar candidate in the Republican Party and also standing for no taxes, the gold standard, and at the same time being more hard-core antiwar and anti–Patriot Act than every Democrat, he broke the link lots of people had in their minds between free-market economics and the rest of the neocon, Bush, big-government conservative agenda."

A "good chunk" is relative, of course, and the fight over progressive minds and hearts on the question of whether Ron Paul is friend or foe continues. That it continues, though, shows that Singleton's friend recognized a phenomenon that lots of progressives have noticed as well—and feel they must combat. Hipster-left sites from Suicide Girls to Mother Jones have felt it necessary to run articles reminding good leftists that they aren't supposed to like Ron Paul—although Ralph Nader has become a full-on Paul friend and was praising him publicly for having a "foundational convergence" with progressives to the *American Conservative* during the 2012 race. Paul himself laid out the grounds for his possible progressive appeal in a March 2011 interview with *Mother Jones*: "As a libertarian, I don't endorse philosophically the many domestic programs and I'm willing to work on a transition. So I say: Let's cut the unnecessary wars. Let's cut the foreign aid. Let's cut all the empire building which costs trillions of dollars and maybe we could tide ourselves over. But for some conservatives to start tinkering with the budget with health care or education for the poor, that doesn't make any political sense to me."

While the *Nation*'s Alexander Cockburn is always a rebel among his fellow progressive leftists, I was amused to find how pro-Paul he was when I interviewed him in 1999. "This is a man who votes against agriculture subsidies in an agricultural district! I will walk ten yards down the political trail with Ron Paul rather than go an inch with [independent progressive] Bernie Sanders," Cockburn said. "I was reading [Paul]'s latest letter to constituents, and he talks of the 'crimes of the U.S. government.' Bernie fucking Sanders never says anything like that." The only part of Paul that Cockburn said he couldn't agree with was that "his reverence for gold is a little excessive." But he added that "if you had twenty Bernie Sanders, it wouldn't be a rump who made a difference. It would just be a Democratic majority." Twenty Ron Pauls, though, could change the whole game in D.C.

Something else besides the foreign policy of peace gave a raffish, leftish tinge to the Ron Paul movement in 2007: the "Revolution" slogan. A logo connecting Ron's name to the concept of revolution, with the "evol" shifted and reversed to make it obvious it was the word *love* backwards. The Lovelution, some began calling it. The seventy-two-year-old obstetrician was running a Lovelution. Why not?

Ernest Hancock was a Libertarian Party activist from Phoenix, Arizona, in his forties. He had watched his beloved Libertarian Party take a turn for what he saw as the right wing and the sellout and needed something new. He had developed that rEVOLution logo for his own 2006 run with the LP for Arizona secretary of state. A peaceful, loving revolution. To this day, lefties on Facebook throw up images of the logo with hand-pumping enthusiasm, unaware of its link with Ron Paul.

Hancock had been itching for a fresh chance to take to a national level his Arizona-based activism, which was always rooted, he says, in "a conscious decision to affect the culture, to penetrate, to inject the libertarian infection into the bone marrow of American politics, for which there is no cure! But to do that we gotta get through the bone." The instant he heard that Ron Paul was planning a presidential run, he saw it as a vehicle for that goal. "You can only do it with a consistent, principled, no-compromise philosophy." Hancock had followed Paul's career enough to know that he was one Republican who had proven he'd deliver a message Hancock approved of.

Hancock called Kent Snyder when he heard the first rumors, to make sure that Ron was really running. "He said yes, and if we get the right level of support, we'll do this, and he was rambling on, and I said, 'Look, all I needed to know was if it was true. Now you'll never talk to me or hear from me again.'"

Hancock knew he had to go guerrilla, outside the sanction of the Ron Paul for President 2008 legal campaign structure. His planned strategy and tactics were not something the campaign would want to officially countenance. And the byzantine complications of campaign finance law made it best that the campaign didn't even really know about his activities, or at least not hear them from him.

That sort of thing makes Ron Paulians wonder where our country has gone wrong. We have this thing called the First Amendment, which declares that freedom of speech shall not be infringed by Congress. When the Supreme Court has decided in the past to ignore this and uphold laws that do just that, the justices have often fallen back on the principle that the key part of free speech, the part that makes it so important to the health of our democratic republic, is the part that has to do with politics, with the free expression of ideas related to governance.

Which makes it a little strange that any attempt to put into action or express one's support for a political candidate is circumscribed by a complicated and picayune set of laws with often severe punishments. This is not only approved of by most "progressive" Americans, but they tend to think those laws are too loose. Now, anything that could be considered directly supporting a candidate for office that was not to be legally treated as an official and thus highly regulated contribution to that campaign needed to be done with no coordination with that campaign.

That suited Hancock and his growing army of national revolutionaries just fine. They didn't want to be managed by any professionals in an office in D.C. They were on the road, in the streets, in the grass roots, on the overpass, clambering up the fence, spreading the message of liberty, peace, and Ron Paul.

Hancock started with some money from friends in Phoenix, rented out an old industrial space, and started making rEVO-Lution signs. The very first one was painted by a friend's son, a twelve-year-old. A revolution of the young.

Hancock's strategy was simple but passionate: encourage and empower interested activists to do whatever they wanted to spread the word about Ron Paul. Hancock believed not just in political liberty writ large but in organizational liberty—in this wired, interconnected world, centrally planning a political activist movement was as foolish and counterproductive as centrally planning an economy. Hancock thought this was a libertarian basic, though he found that the people running Paul's campaign didn't seem to get it.

He had one message for the Ron Paul activists he met across the country: "Decentralize. Do what you want. The answer is always yes. Through all my travels: the answer is always yes. Just do it for yourself, what you want, no committee, no voting to spend other people's money or steer everybody in one direction. That method is broke. It doesn't work. Do what you want, even if others think it's not a good idea: Do your own thing. I did everything I could to break that collective mind-set that you gotta have permission from someone above you to be a good activist. That's bullshit. The only people who really understood were the young people, who were already wired that way through social media."

While he wanted to encourage creativity and self-direction, Hancock did have a simple tactic he trained activists in: get that revolution logo and the name Ron Paul before as many eyes as possible, by whatever means available to political rebels on the cheap. This mostly meant stenciling the "rEVOLution" logo on giant sheets of billboard paper or Tyvek, and getting them hung up anywhere you could drag your body to, ideally places where they wouldn't instantly be pulled down.

The campaign didn't like it at first. "Lew Moore [a campaign director] contacted me and asked that we not use the word *revolution* associated with Ron Paul. They wanted to do it the traditional way, a single type of sign approved by committees and all that bullshit. I told the campaign from the beginning: don't ever think you can tell us what to do." Paul was coming in March to Phoenix for his first campaign stop there, and Hancock recalls the campaign being afraid the media would catch on to his signs and link Paul to them.

Nonetheless, Hancock began traveling the country on his own dime and the dime of his supporters, the Pied Piper of the rEVOLution. "I'd travel to every meetup across the country, and first thing I'd say was: where are the sign makers? There'd be this ocean of blue-haired Republicans and always in a corner somewhere less than a dozen young kids would go, 'Hell, that's why I came, I thought we were making signs, I wanna make signs!'

"Then I'd go over after the meeting to those people and ask, 'Who has a backyard here?' Then for three days we'd make signs. Go to an outdoor ad place and get banners, go to Home Depot or Lowe's and make stencils, get the stencil rolled out, cut it up, and start sign making."

Hancock recalls a love-hate push-pull with the campaign, from people getting kicked out of meetups in Iowa for putting up signs illegally to the campaign beginning to hang them up behind Paul himself on speaking gigs.

Hancock's an anarchist, but he has learned to love the federal highway system for the opportunity to reach a captive audience on the cheap by hanging banners off overpasses. And if the banners get torn down within hours? "So freaking what?" he says. "Two hundred thousand people saw it." And, uh, is any of this illegal? "I don't know," he says. "I don't care." Well, Ron Paul is on record as supporting civil disobedience.

Through all his frustrations with the campaign, Hancock

never regretted Ron Paul. "Never underestimate how much respect I have for his ability to promote the philosophy. He's a thoroughbred horse, and I was going to ride that son of a bitch. I could do it without having to worry about having to make excuses for what he said or meant. It was an enormously freeing scenario, to just do activism and not have to worry about what he's going to say, just make sure he has the opportunity to say it."

After the rEVOLution logo, probably the most iconic object of the Ron Paul 2008 campaign was the blimp. Ah, the Ron Paul blimp. Everyone remembers the blimp. Largest blimp in the world, a Skyship 600, two hundred feet long, seventy feet high, sixty-five feet wide. The only time an airship was used to promote a presidential candidate, as far as anyone can remember. The brainchild of grassroots moneybomb king Trevor Lyman, six hundred thousand dollars raised from the grass roots to see their wild dreams reified in a slow-moving monstrosity in the heavens.

Even the campaign pros who have to admit there wasn't really a lot of *point* to it seem to have a residual, head-shaking affection for the Ron Paul blimp. As Lam recalls, "We knew it was a waste of money, but the XKCD [a popular Internet comic strip that did an absurd sequence mocking what it imagined was happening in the blimp, including Ron Paul himself adventuring about— which he was not] was awesome, and it was memorable, inspiring, a fun thing. I wouldn't say it's a waste, though it was not what the campaign needed [to win]."

When I was doing a dual interview in 2011 with deputy campaign manager Dimitri Kesari and Trygve Olson, a Republican operative who jumped on the campaign in 2011, Kesari ecumenically said a blimp doesn't hurt, but, well, that was six hundred thousand dollars that the campaign didn't have available to spend on things more useful in winning votes. Olson seemed to quietly concur for a moment, but couldn't resist interjecting: "But . . . he had a *blimp!*"

And that sums it up. I mean, who knows what good it did? But it branded the campaign irresistibly as the one with a paradigm-shifting sense of grandiosity and fun. As campaign spokesman Jesse Benton, being as polite and supportive as can be, said to *Politico*: "We're not doing a blimp because traditional political wisdom maybe doesn't say that that's the best way to spend money. But who knows? It could turn out that the blimp is the best thing that anyone's done."

Of course, the media had fun with it, some good-spirited and some mean, though the mean was often pretty funny: Jason Linkins on *Huffington Post* ended his extended head-shake with "Isn't it time to consider whether we might all be better off if we simply gave in to Ron Paul's demands?"

As the blimp took off from North Carolina on December 14, 2007, a crowd gathered, including a man on a horse with a flag at attention and a longhair banging a drum and chanting, "This is what democracy looks like!" *This* was the Ron Paul Revolution! The blimp was emblazoned with "Who is Ron Paul? Google Ron Paul!"

Chris Rye, one of the directors of *For Liberty*, an excellent documentary on the 2008 Paul campaign, got to film in the hangar from which the blimp had been rented. The people running the joint "were insistent that we didn't point the cameras to the other side of the hangar, because these other blimps had secret spy cameras on the bottoms of them and he told us these things were flown over the ocean way up high or over the Midwest or wherever they want."

Paul got booked on *The Tonight Show* on October 30. According to blogger GloZell, a local eccentric who attends every taping of the NBC show, only the lines attracted by Hollywood heartthrobs such as George Clooney, Justin Timberlake, and

Daniel Radcliffe had ever come close to matching the size and enthusiasm of the crowd that began queuing up first thing in the morning, even beating GloZell, who always likes to be first. During the broadcast, Jay Leno respectfully attended to Paul's calls for hard money, withdrawal from Iraq, and a flat income tax of zero. Offstage, Leno got Paul to autograph his copy of the congressman's recent book, *A Foreign Policy of Freedom: Peace, Commerce, and Honest Friendship*. Backstage, fellow guest Tom Cruise came to commiserate with Paul over one of his bills, one restricting the government's ability to force psychoactive drugs on kids. Paul had to ask one of his aides later who he'd been talking to.

During the show, while performing "Anarchy in the U.K." with a reunited Sex Pistols, punk icon Johnny Rotten gave Paul a thumbs-up and a "Hello, Mr. Paul," later adding, "When are we getting out of Iraq?" In between, more ambiguously, he waggled his ass in Paul's general direction.

But Rotten shook hands with the congressman afterward, and according to Paul supporters on the scene he expressed respect to him privately. Paul, watching the broadcast with supporters at a Hollywood Hills fundraiser that evening, shook his head at the aging punk's antics, noting, well, we do promote tolerance . . .

That day encapsulated Paul's surprising campaign. It featured a powerful show of grassroots support, respect from unexpected places, and an infiltration of radical ideas into American mainstream culture. There was the aging iconoclast Rotten, mixing the anarchy he stood for as a kid and the market capitalism he lived out as an adult (the Pistols had reunited to help promote the video game *Guitar Hero III*), symbolizing the range of liberties Paul represents to a movement that includes both Christian homeschoolers and heathen punks. And there was the question so many Americans wanted answered, the question central to Paul's campaign as the only Republican candidate opposed to the war: when are we getting out of Iraq?

Later that night, longtime libertarian movement hand Kerry Welsh was amazed at what he was seeing at that Hollywood Hills fundraiser: "I have attended libertarian, freedom-type events in Los Angeles for the last thirty years, and it's usually the same faces year in and year out. At this Ron Paul fundraiser there were sixty to seventy people, and I did not know *any* of them. They were all new and active libertarian supporters. Someone I spoke with had been a Perot supporter, and by and large these were all people who heard him by accident in the presidential debates and were just transfixed, looked him up on the Internet, and discovered they were natural libertarians."

Nate Howe, a Los Angeles–area computer security worker in the banking industry and an organizer with the local meetup group, was also at that fundraiser, and found "Ron Paul talking to someone who's very accomplished in business and then a kid next to him with a mohawk, and both are saying, 'I like this guy; he's saying go live your life, and if you don't hurt anyone, the government shouldn't bother you.' "

The media began noticing in the second half of 2007 as well. By November, Ron Paul was getting respect from surprising and prominent places. Conservative big thinker George Will called Paul "my man" on ABC. Texas singer-songwriter-novelist Kinky Friedman told CNN's Wolf Blitzer that Paul is "probably telling the truth." Singer-songwriter John Mayer was caught on video informing a pal that "Ron Paul knows the Constitution, and I'm down with that." Even Eleanor Clift, conventional wisdom on the hoof, said on *The McLaughlin Group* that "Ron Paul with his antiwar libertarian message will be the story coming out of New Hampshire for the Republicans."

Still, Paul's media presence didn't always necessarily help him. Maybe it always should have. Many Americans profess disdain for "prepackaged candidates," so Paul should have appealed for his earnest willingness to, media training be damned, treat TV appearances as actual conversations, not just chances to say what he wants to say. On Bill Maher's HBO series in November, Paul patiently explained how he thought the Civil War was a bad idea, and however historically astute, or at least arguable, his reasons were (mainly, that most other industrialized nations managed to eliminate slavery without a massively destructive war), you can't make that a winning position today. However, the Maher crowd did cheer Paul's suggestion that an end to subsidies for Big Oil was a good first step in combating global warming. Small-government radicalism can resonate, when it hits the audience's sweet spot.

The Civil War discussion, which became Maher's running gag for the rest of the night, fed into George Will's take on Ron Paul in *Newsweek*: that Paul's concerns, however brave, are purely anachronistic. The libertarian fight in the American context is already long lost. Paul's belief that the U.S. government was meant to be one of enumerated powers was, Will said, something Paul believed "with more stubbornness than evidence." (Will then went on to quote Madison and the Federalist Papers, generally thought to have some authority when it comes to the Constitution, saying the same thing Paul believes.)

A December 2007 article by David Weigel in the *American Conservative*, a magazine originally founded by Pat Buchanan, summed up well what Paul had come to mean in the lead-up to actual voting in the Republican primaries: "Something is going on," Weigel wrote. "Paul has tapped into a youthful rebellious energy utterly absent from the other GOP campaigns and hardly present among the Democrats. It isn't Right or Left. It's undisciplined. It has its own fringes, which make claims that will never resonate in American politics. But Paul . . . has managed something few expected: he has caught on as the protest candidate, the 'speak truth to power' candidate, the combination Jesse Jackson/ Pat Buchanan candidate, the candidate young people who have not been involved in politics are suddenly talking about."

The campaign relied on local organizations, meetups, and student groups to help it organize local events. Trent Hill, who ran the Ron Paul grassroots effort in Louisiana, remembers a telling moment when the "local leadership" was invited to come take a bow onstage with Paul after a talk, and, well, as Hill recalls, they didn't really think of themselves as having a leadership per se, and everyone is important, so twenty-five people ended up crowding onstage with the candidate. Paul spent the spring, summer, and fall doing these kinds of events, continuing to be largely ignored by the press but building up activist excitement, particularly youth activist excitement, everywhere he went.

The fight in Iowa leading up to the first caucus vote was tough; a major taxpayers group, Iowans for Tax Relief, threw a debate in June and did not invite Paul. Paul fans rented the room right next door and held their own rally at the time of the debate, far better attended and louder.

In the last weekend of October 2007, I tagged along with the Ron Paul road show in Iowa. Over the course of just 24 hours stretched over two days I saw Paul talk to more than five hundred college kids in Ames, more than seven hundred assorted Des Moines citizens, hundreds of state GOP activists, and a few dozen Des Moines–area pastors. I saw a skilled politician with a diverse and disproportionately young band of backers—supporters who stretched far beyond a traditional Republican Party base, who loved their man and his message with an enthusiasm undaunted by whatever his electoral prospects turn out to be.

On the Friday evening before Halloween, Paul was speaking at Iowa State University in Ames. To get from Des Moines to Ames, I hopped on the Constitution Coach, a former school bus owned by Dave Keagle, a Christian homeschooling father of seven. Keagle's wife, Christa, and their children were on board, along with a dozen or so other Paul supporters. The bus was painted red, white, and blue, with slogans summing up Paul's message: "Taxpayer's Best Friend." "No Amnesty." "No NAFTA." "No National ID." "No Patriot Act." "Pro-Gun Owner." "Life." "Liberty." "Freedom." Christa tells me Paul is the first candidate her family has ever been able to get behind 100 percent, with no reservations. She was also impressed with how Paul was able to relate to and remember the names of all her kids on a previous Iowa campaign swing.

I talked on the bus to John Carle Jr., a self-employed CPA in his forties who dabbles in real estate, and his wife, Meredith, a Korean orphan brought to America as a child. Like most of the Paulistas I meet, he's fresh to politics, with no history of activism or enthusiasm for any candidate from any party. He's not a part of any existing Republican base: he's a disaffected independent who thinks he's finally found a politician who "oozes integrity" and "is inspiring the best in people." Paul is the only candidate Carle trusts on post-9/11 civil liberties issues. "If they can pick anyone off the streets and send them to a secret camp," Carle says, "I don't wanna be part of that country."

The talk at Ames drew an overflow crowd. There were a few longhairs, a few punks, but it was overwhelmingly a conventional gang of well-groomed midwestern youth who happened to be wearing "Ron Paul Revolution" T-shirts. The event got no free local or campus press. The crowd was gathered almost entirely through Meetup and Facebook.

"I hear you've got a revolution going on," Paul began by saying, "and it's being led by the young people." Then he recited his first big applause line: he's not much for passing laws, but he might consider one requiring the next election to be held on the Internet.

Those were the only explicit nods to the crowd's youth and online activity. From there on, it was all classic Ron Paul: Get rid of the income tax and replace it with nothing; find the money to support those dependent on Social Security and Medicare by shutting down the worldwide empire, while giving the young a path out of the programs; no draft; and a foreign policy of friendship and trade, not wars and subsidies. He attacked the drug war, condemning the idea of arresting people who have never harmed anyone else's person or property. He stressed the disproportionate and unfair treatment minorities get from drug law enforcement. One of his biggest applause lines, to my astonishment, involved getting rid of the Federal Reserve. Kids had gathered not just from Iowa but also Wisconsin and Nebraska, taking a classic hop-in-the-van college road trip, to hear a seventy-two-year-old gynecologist talk about monetary policy.

He wrapped up the speech with three things he doesn't want to do that sum up the Ron Paul message. First: "I don't want to run your life. We all have different values. I wouldn't know how to do it, I don't have the authority under the Constitution, and I don't have the moral right." Second: "I don't want to run the economy. People run the economy in a free society." And third: "I don't want to run the world. . . . We don't need to be imposing ourselves around the world."

Paul did not mention abortion or immigration—areas where his views are more conventionally conservative and not of great appeal to this age group. Paul's style of libertarianism includes a populist streak of distrust for foreign forces overwhelming our sovereignty, whether through the United Nations, international trade pacts, immigration, or a feared "North American Union" between the United States, Canada, and Mexico, but he was not talking about it that day in Ames.

On the ride back to Des Moines, I met, among other Paul fans, Bryan Butcher, a fifty-year-old high school teacher and part-time drummer for a belly dancing troupe. He's a ponytailed former marine who had thought of himself as a "social liberal" and an Obama fan. "I feel we do need to take care of people," Butcher said. But Ron Paul has helped him see that "the socialist idea of government taking care of people hasn't helped, that *people* need to take care of people, and that's the smart way to go."

The Paulistas delight in their independence and fervor. At a press conference after the Ames talk, a *Pittsburgh Tribune-Review* reporter asked the candidate about all the Paul signs he had been seeing around Pittsburgh. "You guys must have a big operation there," he said.

"If we do," Paul said with a small smile, "we don't know about it."

The next night I was on Des Moines's downtown drinking strip after Paul had spoken at a state GOP dinner. I was sitting with two Paul staffers and two Paul fans. A tipsy young Romney supporter approached us. She actually liked Ron Paul, she granted. She could even call him her second choice. But Ron Paul fans? They're outside agitators, she insisted, almost scary in their intensity. Iowans don't appreciate their shouting, chanting style of campaigning, or their insistence on sticking their huge, silly "Ron Paul Revolution" signs in places they do *not* belong, often violating both propriety and the law.

Both the official campaign and rowdy out-of-state grassroots volunteers worked very hard for the Iowa caucus for Ron. He came in a disappointing fifth in the Ames straw poll, first bellwether of where the state would go, as evangelical Mike Huckabee began his rise. Fans across the country were doing phone banking for Paul's campaign, calling Iowa voters; in Los Angeles they were mailing handwritten letters to likely Paul voters among Iowa's Republicans. For the endgame, the campaign sponsored hundreds of college kid volunteers to ship in to work getting out the vote in the Iowa chill. Jeff Frazee, who now runs Young Americans for Liberty, was wrangling the volunteers; he remembers sending these kids, many from warmer climes and not properly prepared, out to door-knock, and counts it as a triumph that no one ended up hospitalized, in a car wreck, or dead. For a week after the race had ended in Iowa, Frazee was recuperating—he got terribly ill toward the end—and waking up every morning imagining the room was still filled with a gang of Ron Paul kids sleeping on his floor awaiting instructions for the day.

At the very end of the Iowa process, one of those weird misfortunes that made some Paul insiders believe quite intensely, but without sure proof, that they were being targeted by enemy provocateurs happened: the guy in charge of putting together a usable database of known supporters' names, addresses, and phone numbers for get-out-the-vote calls on election day disappeared, and young volunteers had to try to re-create it in a forty-eight-hour sleepless frenzy of data entry.

Paul came in a very disappointing fifth in the Iowa caucus on January 3, 2008. For those who learned to love him because of the Giuliani dustup, they could take some pleasure that Paul got nearly three times the number of votes Giuliani did. Giuliani's people tried to spin it later by claiming vaguely that their man had not really been competing very hard; yet he'd made more campaign appearances in Iowa leading up to the caucus than Paul did.

New Hampshire was the next frontier. The delightful documentary *For Liberty*, made by Paul activists Chris Rye and Corey Kealiher, includes clips of amped-up Paulistas braving the insane weather, laughing atop enormous snowbanks, cracking each other up—"We don't do it because it's easy, we do it because it's hard"—in an absurd New England accent. "One small step for man, one giant leap for Ron Paul," as they laugh and leap. You see activists trying to spray-paint signs in the freezing cold, the can freezing up, while the activists admit, well, my hands are too numb to even know it is cold.

Vijay Boyapati was a computer programmer from Australia working for Google. He had gotten interested in libertarian politics via the Internet, and was so excited at the thought of hearing directly from Ron Paul that when he heard the congressman would be speaking on Google's own famed Googleplex campus in Mountain View, California, in June 2007, he knew he had to attend—even though he was based in Washington state.

The Google appearance was a pivot point in Paul lore, helping him break through to a new level of respect with a savvy, well-connected, well-heeled audience. (As usual, it resulted in a much-watched and forwarded YouTube video of Paul's talk.) Boyapati came up to Paul afterward and handed him a check for the maximum legal contribution, telling him, "I wish I could write a blank check." Boyapati's enthusiasm was nonetheless probably the biggest payoff from that Google appearance.

In a conversion experience not exactly common but also not unknown in the world of Ron Paul, Boyapati decided to quit his highly coveted job; start his own political action committee, called Live Free or Die; and move to New Hampshire. And try to get a lot of other people to follow in his footsteps, by renting a bunch of houses and apartments there, to shelter whatever scruffy Roniac wanted to come spend the winter trying to win one for the revolution in the Granite State.

Boyapati raised and spent $140,000 toward this effort. He helped organize a Martin Luther King Jr. Day moneybomb that brought in nearly $2 million. "I bought literally every air mattress in New Hampshire. I'd go to Wal-Mart and Target and empty the shelves into a shopping cart." He rented about fourteen houses for canvassing volunteers. He assigned to each a house captain responsible for getting his troops out and ideally instructing them on how to do effective door-to-door campaigning. He brought together the usual Paul fan motley: concerned veterans, pierced anarchists, conservative Christian moms, real estate brokers and homeschoolers and weapons enthusiasts and peace hippies. One Paul supporter bought and shipped up to New Hampshire twenty thousand copies of the U.S. Constitution. "We'd go out and knock on doors with these," Boyapati says, "and the pitch would be, 'This is the U.S. Constitution; have you read it, and do you know whether any politician is consistent with it?'"

Sleeping on strangers' floors crammed with dozens of other strangers; knocking on the doors of countless even stranger strangers—this was all made more difficult by the fact that this was New Hampshire and it was the heart of winter. Often the Paul troops would hear warnings that it was dangerous to have any of your skin exposed to the elements for even a moment or two; but even on days like that they could get a hundred or more people so dedicated to Paul that they'd dare arctic-like winds to wave signs or knock on doors or march or spontaneously protest Fox debates from which Ron had been excluded. All to share the good word: freedom was possible, and freedom was good. Not everyone wanted to hear it, or at least they didn't want to hear it in connection with this peculiar Ron Paul character who all too many New Hampshirites visited by Operation: Live Free or Die knew was too crazy, or unelectable even if they personally admired him.

All the Paulites learned to enjoy each other, mostly, and to enjoy being in casual, chosen communion with all types of strange people, the likes of which they'd never otherwise meet. Some of them would casually carry around a shotgun, and happily explain gun use and gun safety to the others; others would sing freedom folk songs late into the night. All for Ron Paul.

New Hampshire had a built-in advantage for the Ron Paul cause. The Free State Project, a libertarian plan to move thousands of like-minded liberty lovers to the state to help shape its politics and culture in a libertarian direction, had been swinging there since 2003 already. People in the Free State circle were key volunteers for Ron Paul, and the Paul effort attracted new people to the Free State; some of them stayed or returned later to stay.

Boyapati learned lessons the government loves to see grassroots activists discover when navigating the field of campaign finance law: amateur political action is onerous and you are better off not getting involved. "The Federal Election Commission is even more intrusive and Orwellian than the IRS," he says. "They are very particular about how reports should be filed, and if it's not done in the exact right way you get in trouble. They were harassing me for a while about some reports that didn't have a couple of receipts, for like one hundred dollars out of over one hundred thousand dollars, and that was a painful thing to go through.

"The FEC absolutely discourages political involvement and kills a lot of small grassroots efforts, but it's advantageous to these huge corporate PACs. If you're a corporation creating a PAC then you have the money to hire lawyers and accountants; but doing it like I did, with all that red tape and bureaucracy and figuring out arcane software for filing reports—you have to figure all that out or go to jail. I didn't know better then, but I definitely wouldn't do something like that unless I could grow the PAC big enough to hire its own lawyers."

Talking to Paul activists about the salad days of 2007–2008 is fun; you run into very little regret, even in those who wouldn't dream of doing the same thing again. Brinck Slattery, who helped run the New Hampshire operation for Paul, remembers a mob of Paulites three hundred strong following Paul as he left a July 2007 ABC interview in New York with George Stephanopoulos, the supporters chanting and yelling and following him into Grand Central Terminal. "Amtrak security thought a riot was happening. It was something else, really strange and not replicable," he says, clearly still buoyed by the memory. That same week, as Paul hit various high-profile New York City media, a supporter threw him a fundraiser in a former S&M dungeon.

Slattery slotted into a post-moneybomb campaign where he recalls money being no object. "Anything I asked for I would get. They were very generous with employees, reimbursed, paid for my housing, a lot of generous good stuff. More money than they knew what to do with. We'd wrap cars with Ron Paul's face, print out our own homebrew signs. Since I flipped rapidly from grassroots to official campaign, I missed the discontent between the grass roots and the campaign. I heard people complain they weren't getting resources to people that wanted them, but in my experience working for them we never had a problem. Volunteers kind of became the problem. Some were overenthusiastic, arguing with people at the mall, getting in fights with people about open carrying of guns, a bunch of stuff that made locals wary. Putting signs on people's private property when it hadn't been okayed, getting locals pissed off."

Paul's 2012 political director Jesse Benton noted something peculiar about Paul volunteers compared to those in a more typical Republican campaign. "Volunteers in a traditional Republican campaign want to be part of the party system," Benton says. "So if you give them goals, like make a hundred phone calls or recruit five people," he says, "you can generally count on them to do it because they are there to help them move up in the party hierarchy. Our people are passionate, but they might want to wave signs or go to fairs, they don't necessarily want to sit and do phone calls or do things that seem more like drudgery. And we don't have the traditional stick of, 'Hey, do you want to move up in the party establishment, get in the good book of the local GOP?' That doesn't motivate Ron Paul people."

As Jared Chicoine, in charge of New Hampshire for Paul in both 2008 and 2012, stressed to me over and over, phone banking is the heart of an insurgent campaign that isn't flooding the media with ads—yes, phone banking, not chatting about Ron Paul on the Net, flooding cable news channels' online polls, or doing sign-waving and Constitution distribution in painful, murderous cold. You need to make tons of calls to identify your people, your possible people, and those who aren't worth bothering with. And you need to find out what motivates your people and your possibles, and follow up with them in ways targeted to their interests. And once you have solid lists of your people and their concerns, you need to make sure they actually come out to vote when you need them to. That adds up to lots and lots of phone calls. "No one in their right mind enjoys coming into an office and making cold calls to strangers for three hours," Benton admits. "It's a terrible job, but it's the single most important thing done by grass roots in a campaign." The Paul campaign by the 2012 election season had a sophisticated and effective system that allowed Paul fans from all over the country to call wherever the campaign needed them to call as any given caucus or primary approached.

Phone banking is a time when message discipline is important. But as Slattery remembers, every volunteer "had his own vision of who Ron Paul is that they are going to want to promote. I remember I had this guy who kept coming in with his own scripts for phone calls, 'Let me do this awesome rap about the Bilderbergers and the Trilateral Commission forcing public education on us.' Uh, that's not what we need. And from another direction, one guy printed out his own flyers of Ron Paul's face superimposed on a weed leaf: 'Ron Paul: Free the Weed.' Then he handed it to Ron Paul's granddaughter. We got yelled at for it even existing, but this guy drove up from North Carolina and had printed up his own pamphlets. It's not something we could control."

Whether it's something they should even try to control is a question that still divides the Ron Paul world. Remember Ernie Hancock, who seeded the undoubtedly successful rEVOLution banner meme with his resolute the-campaign-can't-tell-me-what-to-do attitude? As with any ultimately unsuccessful endeavor (in terms of winning votes), the world around the 2008 Paul presidential run swirled with bitter second-guessings, mutual recriminations, deep misgivings, and lots and lots of complaints about poorly planned, poorly executed, and mistakenly conceived efforts on the parts of everyone but the complainer.

Not every press call was returned, or every press conference properly announced to the press; the paid media ads were an embarrassment to most of the grassroots and even many official campaign workers. No one was an experienced expert on the very arcane science of GOP delegate allocation in caucus states, each of whom has its own rules pretty much designed to make things hard on confused amateur outsiders like the Paul campaign, a campaign that was disconnected from the existing Republican Party machinery. Ballot access signature collection was similarly a confused mess, with grassroots workers often swearing that only they made up for campaign workers letting things fall through the cracks. And the official campaign had complaints about the grass roots (though it should be stressed they all agreed that on balance it was a great thing Ron had the enthusiastic fans he had), similar to the ones Slattery had above, to match every one the grass roots had about the pros: they could be an undisciplined embarrassment at times. (However, everyone has their unexpected stories of a Ron Paul fan's grassroots charm, like a guy Neal Conner found while himself out doing Ron Paul activism in rural New Hampshire, who was just giving people rides in a balloon he'd festooned with Ron Paul logos and messages.)

All of this effort and conflict and tumult were supposed to have its big payoff on January 8, the New Hampshire primary. New Hampshire, the live free or die state. New Hampshire, the state that had four sitting Libertarian Party members in its legislature in the 1990s. New Hampshire, the state chosen by the members of the Free State Project.

New Hampshire, the state where Ron Paul came in fifth, with 8 percent of the vote. Messaging or name awareness seems to have been a problem—polls show, crazily, that he only pulled 16 percent of those opposed to the Iraq War, even though he was the only Republican candidate who was against it.

What went wrong? There is no reliable polling on the question of why one did or didn't vote for Ron Paul. And while the fans were discouraged, many to tears, the vote total came in pretty much as the campaign's polling suggested; they were not surprised. While Zogby before the race was saying Paul could pull up to 17 percent, Chicoine says, "I knew what was going to happen an hour before the polls closed, three hours before the polls closed. We were running robo-trackers all day, and I knew. . . . I could have told you the day of, it's going to be between eight and ten percent."

My then-colleague at *Reason* magazine David Weigel, who now covers the American right for *Slate*, was in New Hampshire before and during the election, and wrote about it at the time: "Ron Paul voters and volunteers, men and women, pinching their eyelids and daubing their tears in . . . crushing disappointment. . . . [It was] a terrible night for Ron Paul. The theory that Paul could perform well in New Hampshire has been shredded, as has the theory that an amorphous Ron Paul vote was not being counted by polls." Some fans insisted it was a Diebold voting machine conspiracy and urged the campaign to fight for a recount; the campaign, knowing better, did not want to fight. (Another candidate did demand and get a recount, in which Paul got . . . 38 more votes.)

Weigel then, and other observers then and since, saw three causes of Paul's failure: the message, the ads, and the fans.

The message. One should never forget this when thinking about the Paul movement—especially if one is *part* of the Paul movement: most Americans just don't agree with Paul's particular views. They don't agree that the government's only purpose is to protect our individual right to be protected from force, violence, and theft, and that the federal government's power was even more circumscribed than *that*.

The ads. Following what seemed to be a combination of the

logic that Ron Paul's true people already knew him and could be expected to vote for him, and that such people were insufficient to win a Republican primary, the official campaign ads tried to sell not Ron Paul in all his Ron Paulness, but a particular skewed and right-wing-friendly version of something that, if you squinted, was like a part of Ron Paul—though never the Ron Paul that Ron Paul himself chose to project in his own speeches. They emphasized abortion, immigration, and the general notion that even if you disagreed with him, he still had *some* good ideas, and was "catching on, I'm tellin' ya!" as one notoriously amateur New Hampshire TV commercial claimed. As Benton says, "We are not running for mayor of *The Daily Paul* [a prominent Paul grassroots fan website]. Ron is running for president, and we do research, we have numbers, and with the ads we are trying to do the best we can with our resources we have." Benton says that the campaign ads' critiques from the Paul grass roots "need to understand these commercials are not for them" but are for the campaign's best understanding about the Republican voters they need to convince.

The fans. This touches more on that most poignant conflict in the world of Paulian politics, discussed above: the conflict between the campaign and its grassroots activists. The quickest summation, which I've heard from both official, paid campaign staffers and grassroots enthusiasts, is that too many of the things that the excited fan wants to do actually do nothing to convince a primary voter to vote for Paul—and in many cases the result is less than nothing. Chicoine and I had almost a running gag going in our interview—whenever I'd tell any story involving a grassroots activist, he'd respond with "I'd rather they were making phone calls."

But the fans' hearts were on fire; Dr. Paul had cured their apathy. They'd join together to place ads in the Ames paper on straw poll day, an image of Ron Paul's face formed of the tiny images

of Ron Paul fans from all over. (A fan went through the trouble
of forming a legal PAC just to make sure they were on the right
side of the law on this. Remember, in the land of the free and the
First Amendment you can and will be punished for what you say
about a politician!)

They'd make movies; Chris Rye and Corey Kealiher were
amazed at all the interesting grassroots furor they saw and decided to start filming. They drove around in a 1985 Mercedes
300 turbo diesel that they converted to run on waste vegetable
oil; their jerry-rigged filters were often taxed by the goopy quality of the used oil they were getting for free, and by the end the
filters failed entirely; they had to drive home from California on
regular, paid diesel.

Fans loved to stand on the side of the road waving signs, telling
people to Google Ron Paul! YouTube Ron Paul! They loved making viral videos. They loved answering any negative online article
about Ron with lots and lots and lots of comment thread retorts.
They liked arguing with each other and skylarking campaign plans
on online forums like DailyPaul.com and RonPaulForums.com.

Slattery himself, official campaign worker in New Hampshire, still had much love for his old-school activist days; he remembers fondly how in his New York- and Connecticut-based
meetup group, they were filled with the thrill of a fresh world
where everything was an adventure and no one was there to tell
them they were doing something wrong. He recalls a quick en-
velope pass during his group's first meeting that gathered over a
thousand dollars.

They decided it would be cool if they had some Ron Paul
signs, so, wham, they had some money and then they had some
signs, purchased from an online Ron Paul swag vendor. "We did
stuff like that constantly in our New York City group. A lot of
spontaneous grassroots enthusiasm on our own time, spending
time and effort to help out this campaign that wasn't even re-

ally like a political campaign to us. At the meetup stage we were not concerned with being electorally effective at all. It was just, let's get this dude's name everywhere, get banners with his face hanging everywhere, let's see his name spelled out from Google Earth!"

Mitch LeClair of the University of South Dakota told me how he decided to express his enthusiasm for Ron, to spread the message, in the medium most available to him: the snow that covered the grounds of his campus in a flawless bed of undisturbed white. Snow-angel style, Mitch spelled out: GOOGLE RON PAUL.

When he woke up in the morning, one of his fellow students had had his way with LeClair's handiwork. He gazed out his dorm room window to see his fellow students being instructed in inscriptions in the snow: GOOGLE RON PAUL'S PENIS.

S omething else had happened right before the New Hampshire vote: the *New Republic* rediscovered the nasty newsletters that had arisen to haunt Paul in his 1996 congressional campaign. In an article by James Kirchick, "Angry White Man," Kirchick found and presented more unpleasant race-tinged jabs from the early 1990s *Survival Report* and *Political Report* newsletters, including at Martin Luther King Jr. for being a philandering communist, and at the blacks of Los Angeles for allegedly only ceasing to riot after the Rodney King verdict when their welfare checks were ready, and at blacks in general for being prone to carjacking. They also included dumb comments about AIDS that showed, at the very least, extreme insensitivity and ignorance about the matter, including that it could be transmitted through saliva.

The story actually got more play among Paul's fans and the specifically libertarian world than it did outside, though when pressed about it on CNN Paul got uncharacteristically passionate in his objections, while claiming to have no idea what ghostwriter might have been responsible. Paul told Wolf Blitzer in January 2008 that

> everyone knows in my district I didn't write them and don't speak like that . . . everybody knows that I don't participate in that type of language. . . . You are really saying, "Are you a racist?" and I'm not. Rosa Parks is one of my heroes, Martin Luther King is a hero, because they practiced the libertarian principles of civil disobedience and nonviolence . . . a civil libertarian like myself sees everybody as important individuals, it's not the color of skin that's important; like Martin Luther King said, what is important is the character of the individual. . . . I attack two wars that blacks are suffering from, one war overseas, in all wars minorities suffer the most . . . and what about the war on drugs? What other candidate will stand up . . . and say they will pardon all blacks, all whites, everyone convicted for nonviolent drug acts?

What does this scandal mean in the big picture of Ron Paul? Its very existence is a burden on the soul of Paul's enthusiastic fans—it's a spirit and tone that has nothing at all to do with the Paul they've met since 2007. That sort of rhetoric is part of a small slice of Paul's history, which lends credence to his later assertion that he did not write them. No one knowledgeable about the world of big-name newsletter writing would assume that the man whose name is on it wrote it.

That doesn't matter to many of his critics. Ghostwritten or not, the writings came out under his name, and presumably he knew about them, though even that isn't certain—if he had grown to trust certain writers and editors it's certainly conceivable the full-time doctor wasn't even vetting it before publication, as he said he wasn't. He also did not instantly disavow having written it when the story first arose in 1996. The denials came

in 2001, during an interview for *Texas Monthly.* The charitable
will interpret that as him doing the honorable thing initially and
taking responsibility for what went out under his name. The un-
charitable will assume he was lying to cover his ass later. (One
fan musing on the matter on the Internet noted that Paul chose a
time when the incident *didn't* matter to his political career to say
he didn't write those things; that lent credence to the notion that
it wasn't a mere political face-save.)

Most Paul fans are happy to just accept that he didn't write
them, that most of it is at the level of crude right-wing talk radio
shock-comedy of the period as opposed to virulent racism, that
he's said and done nothing since and certainly not in his career as
a politician or presidential candidate that shows the slightest hint
of racism (or any tendency to speak in such crude and childish
terms about any issue, ever, anywhere). They will also point out,
as Paul himself has, that policies Paul is pretty much alone in advocating in public life, such as drug legalization, would do more
to help more minorities in this country than those of any other
politician, since these minorities are disproportionately targeted
by drug law enforcement.

The sin Paul can be accused of most justly is something he
has also mostly been innocent of: pandering. He or the people
making decisions about how to frame and sell him in these early
1990s, a time when he was free of the explicit responsibilities of
holding office, decided to go straight for the most resentful and
ugly end of the American populist right, the stereotypical angry
white man, resentful and crass, willing to believe and play on ste-
reotypes about black people (they are prone to crime, fleet-footed,
a looming danger to the white man's way of life).

The newsletters are a political tragedy of sorts: in a misguided
attempt to reach out to one small subset of the American political
scene by speaking the language of their resentments, the people
around Paul severely damaged his chances of reaching out to a far

more significant constituency today, especially a progressive left disgusted with war and corporatism and rampant assaults on civil liberties. Those types, while they do not love free markets in principle, might hate them less than they hate the fascistic corporatist alliance that has been beggaring America of late, in ways more and more obvious since Paul's first presidential run. A public representative of this mind-set, *Rolling Stone* reporter Matt Taibbi, has openly said the newsletters are what make it impossible for him to be publicly supportive of Paul, though he agrees with him on many things.

Paul fans get very angry when the newsletters are brought up. For complicated reasons involving superego and self-image, it's actually more likely that libertarians who might fear being tarred with the newsletters themselves feel it necessary to harp on them in order to disavow them, since they otherwise seem to think like Paul. How important it will prove to be to the Ron Paul story (it barely hurt him in 2008—the story simply didn't get that much play outside the world of people who already cared deeply about Ron Paul) remains to be seen, though it is a shadow that haunts him, erupted again loudly in the weeks before the Iowa caucus in 2012, and continues to haunt the people responsible for his campaign and image. I personally know or have met more than a handful of people who refuse to give Paul any credence as a positive political force because of them, and there are surely many more.

Unnamed editorial writers at the *American Conservative,* a magazine cofounded by Pat Buchanan that eventually endorsed Paul in 2008, wrote, "The authors of Paul's newsletters may be the first people in history who secretly wanted to write about monetary policy, but concealed their true selves by pretending to be racists."

After Ron Paul failed to take off in Iowa and New Hampshire, Justine Lam recalls that the campaign strategy heading for

Super Tuesday and beyond involved concentrating on caucus states rather than primaries, since in the former, delegates could in theory be piled up without actual huge statewide wins. (In a nutshell, you can win delegates in caucus states without actually getting a ton of citizens to vote for you.) But after Super Tuesday, Lam thought it was clearly hopeless and hoped Paul could admit as much and loose his supporters' energy and enthusiasm to flow elsewhere.

Instead Paul issued a somewhat hot-and-cold statement to the effect that they'd be making the presidential campaign leaner and tighter but would still fight on. However, he also had to mind his congressional race, where he faced a primary challenge from Chris Peden. Peden was actually beating Paul in some early polls but ended up crushed, 70–30. Many staffers were let go, but the campaign crawled on. "We tried to convince Ron to make a video thanking people and saying it's over," Lam says, "but Ron could never say it was over."

In early March he released a video with a mixed message. Paul talked up a continuing grassroots effort to promote liberty, with emphasis on "the message of human liberty," which, he said, was explained at greater length in his new book, *The Revolution* (which became a *New York Times* number-one bestseller). In the video, he talked through the basic shape of the Paul movement strategy that went into effect after 2008, including hyping his LibertyPAC, which helped fund libertarian-leaning campaigns. He also pointed out that philosophy and ideas were as important to politics as candidates and elections. He said that "the presidential campaign will soon wind down" while also calling on his supporters to continue trying to rack up delegates in the caucus states and promising he'd continue to hit the campaign trail where he was wanted. It was an ambiguous message that confused many of his supporters, but he tried to spur them to continued action by saying, "I don't mind playing a key role in the revolution, but it has to be more than a Ron Paul revolution. I have always claimed this whole effort was much bigger than one individual."

One campaign worker in 2008 says that "after Super Tuesday it was apparent Ron wasn't going to win, and there were competing opinions amongst the grass roots and campaign and within the grass roots and campaign and Paul's family about whether to continue or shut down. Ron by nature is not going to be an authoritarian type and make hard decisions that will upset lots of people close to him." When it comes to things like personnel and managing his own operations, "he can go with the flow in a way, and I think Carol wanted to keep the campaign going, and that has more influence on Ron than anything."

Lam recalls morale collapsing among the rump staff after Super Tuesday: much Guitar Hero was played; Snyder got ill with the pneumonia that killed him (this caused a mini-scandal later for Paul when it became public that the campaign had supplied no insurance for Snyder and he died leaving behind four hundred thousand dollars' worth of bills for his family; Paul fans did launch a national campaign to raise funds to help defray his family's expenses). Snyder had, Lam recalls, a multifaceted plan to take the revolution forward organizationally, some of which came to fruition with things like Campaign for Liberty and Young Americans for Liberty. She recalls he envisioned a legal action group that would pursue liberty in the courts as well.

Lam got a fair amount of press hype for being the queen of Ron Paul's online presence, but she left the Ron Paul world behind with no regrets toward the end of the 2008 campaign, which end she felt was unnecessarily dragged out. She doesn't tend to bring up Ron Paul much anymore in conversations with civilians—"Why destroy the mood?"

One could write an entire book about how delegates are assigned in the Republican primary process, but it would be, I assure you, a confusing and uninteresting tale. In Nevada, all delegates are assigned through a convoluted process at a convention; the Paul forces were clever enough and organized enough at the lower level that they were primed to select more delegates than their raw numbers might have indicated. State party officials, not wanting to have to send a mostly Ron Paul crew to the national convention, illegitimately shut down the entire convention before the delegate selection process could finish. "I've heard of a faction leaving a convention, but never a convention leaving a faction," says Jeff Greenspan, the Paul campaign's caucus delegate wizard for the West. Legal challenges ensued, and Paulites eventually won a 15 percent representation among the delegates, chosen by the national party.

The Republican Party showed its disdain for Paul and his fans nearly every chance it got. Saul Anuzis, the chair of the Michigan Republican Party, openly called for Paul to be kept out of GOP presidential debates. But Republican Ron Paul remained, despite rumors (and the hopes of some of his fans) that he'd launch a third-party challenge.

Ernie Hancock was invited to a Paul strategy meeting in March featuring the stars of both the official campaign and the grass roots, when it was clear the presidential race was winding down. Hancock got wind then of the Campaign for Liberty strategy, the idea of working forward within an explicitly political—and in his reading explicitly Republican Party—mode. Hancock wanted nothing to do with it. "That was their rhetoric, 'How can we save the Republican Party?' to which I responded, 'Save it? I'm looking for a wooden stake and mallet!' What the hell do I wanna save the Republican Party for? But that was their goal."

Hancock saw this CFL approach as merely a way for the Paul machine to stay connected with a set of fundraising sources more comfortable working in a standard political context, which Hancock didn't care about; he cared about grassroots activism and spreading the ideas of liberty mind-to-mind, not electing candidates per se. All that Paul's campaign pros wanted to know from *him*, Hancock thought, was not the methods of successful rEVOLution in the decentralized streets, but just his fundraising secrets. He didn't have any. The Paul professionals, in his estimation, "didn't understand spontaneous order. It doesn't exist, in their heads."

The run for president was pretty much over, though Paul continued to do personal appearances on college campuses that still drew huge and excitable crowds. As Lam says, Paul "was like a little kid in a way, the way he loved the message, researching it, reading papers on it, talking about it in intellectual discussion, talking to university students. He just loved talking about these ideas and that is when he glowed—around students. But a lot of time on the basic campaign grind, he seemed grumpy and stressed out. He didn't love being pushed out all the time, and overscheduled; he liked time for exercise, his routine, his family."

In June, Paul made it official. The $3.5 million left was rolled over to his congressional campaign committee. But Paul announced the next step: Campaign for Liberty, a national grassroots legislative pressure and education group for Ron Paul politics. It would make its public debut in a convention to be held across town from the official Republican one. This Paulite convention would be called the Rally for the Republic.

The revolution, though, still wanted to keep running. Or walking, or otherwise moving to prove their devotion to the cause of Ron Paul. People liked to walk for Ron. Or march, or bicycle, or drive, or otherwise transport their bodies in colorful and attention-getting ways to show that, dammit, they had the power to move themselves for Ron. Maybe they did not have the power to make him president. Maybe they did not have the power to bring liberty in our time. But they could put one foot in front of the other for their own reasons. As Iggy Pop sang (in a song I first heard as a cover by my old college roommate, who first told me about Ron Paul in 1987), any old time they had a right to move. Any way they want. Any old time.

In the height of July summer heat in 2008, Paul fans threw a Revolution March in D.C. that drew many thousands, marching to the west lawn of the Capitol. Paul grassroots activist Gary Franchi, who helped organize and emcee the event, remembers it as "a magical and beautiful moment. In all my years of activism, that is the pinnacle moment that I will share with my children, my grandchildren. I will show them pictures of me advocating for freedom and revolution at the United States Capitol."

Paul fans wanted to walk across something more impressive than just D.C. Michael Maresco, a Hawaiian stage rigger, had walked across Oahu spreading the word of Ron Paul. He was inspired by Kelly Halldorsen, a homeschooling mother who drives around the country with her family in the "Un-School Bus," who got press for Ron by walking solo thirty-eight miles from Dover to Concord, New Hampshire. She did it because she was annoyed that George Stephanopoulos told Ron Paul to his face that he couldn't win. She thought that was disrespectful not just to the man but to the ideas he represented. She did her march with no preparation, and lost a toenail. But Ron Paul fans gave her much love and support, and Ron even called her along her way. Emboldened, she did the same thing in Florida, this time walking fifty miles from Boca Raton to Miami.

Maresco decided walking wasn't enough and so set out to become the Ron Paul rider—bicycling from the Santa Monica Pier in California to the Washington Monument. He'd crash most nights along the path with Ron Paul buddies he'd made through the meetups, and stop most days to hand out Ron Paul literature and talk up Ron Paul at gas stations and convenience stores. He too started cold, with no long-distance bike riding experience. Why? People needed to know about Ron Paul.

Leading up to the Rally for the Republic, a group journeyed from Green Bay, Wisconsin, to Minneapolis, walking, walking all the way, stopping to hand out Ron Paul literature and talk about the Constitution and Ron Paul and liberty, and one of them carried a flag on a sixteen-foot pole the entire time. Paul himself joined them in that last stretch in Minneapolis. When they got to the Federal Reserve building, Ron jumped up on a little ledge to give a speech and a fan slapped a Campaign for Liberty sticker on the Federal Reserve emblem on the side of the edifice.

When they finally got to the Target Center, the end of their journey, they melted into a huge crowd of Paul people. Many of them knew of the group's journey, some of them had supported the walkers through online donations. The walkers were taken in and acknowledged and then they joined the bigger revolution. Filmmaker Chris Rye, who did the walk from Green Bay, had his own PAC buy Ron Paul billboards around the city. One on top of the First Avenue nightclub, where Prince did his thing in *Purple Rain*, had a popular declaration from Ron's number-one bestseller *The Revolution*, "Truth Is Treason in the Empire of Lies," along with a picture of the book. Another featured a quote from Paul's book, about how we are building bridges overseas while ours here are falling down—only a year earlier a bridge between Minneapolis and St. Paul had indeed fallen.

Rye went to shoot at a Ron Paul book signing at Borders. "It was the longest line of people I'd ever seen," filling the mall from end to end. "He comes in, this little old guy in his yellow shirt, and waves, and the whole crowd goes wild. Random shoppers stare in wonder—who *is* this about?" Just some guy who couldn't get anywhere in the Republican presidential primary. . . .

Trygve Olson, a Republican operative working in 2011 for the Paul campaign, was just another Republican at the Republican convention in 2008. The night of the Rally for the Republic, he recalls, "I happened to be across the street at the First Avenue club, where Sammy Hagar was having a concert. I was not happy with Ron Paul people at that point, because I grew up in the 1980s and I was excited to see Hagar, and I didn't calculate the drive time with fifteen thousand Ron Paul fans all packed outside the Target Center across the street."

The rally demonstrated vividly that this was more than just another of the hundreds of losing presidential campaigns in American history. When does a candidate drubbed in the primaries by his party have the will, the organization, the savvy, and the sheer cojones to plan and execute a counterconvention across town from his party's nominating convention, and draw nearly fifteen thousand people? This was not the end of anything; this was the continuation of a revolution that began within the Republican Party but was bigger and wilder than the Grand Old Party could sustain.

It had Ron Paul singers and Ron Paul intellectuals and Ron Paul economists and Ron Paul celebrities and, most of all, it had Ron Paul. TV conservative Tucker Carlson emceed most of the night. He bailed out when he found former Minnesota governor and now popular conspiracy theorist Jesse Ventura's references to 9/11 conspiracy ideas in his speech—something he didn't care to be associated with. He didn't want to make a big deal out of it or embarrass Paul or even Ventura, Carlson says; he just didn't want to have anything to do with it anymore.

Those who were there remember the rally as an unalloyed, joyful triumph, the natural continuation of something none of them wanted to give up on. One of their own, Adam Kokesh, a Veteran for Ron Paul, had gotten kicked out of the Republican National Convention for confronting McCain on the floor. He arrived at the rally a conquering hero.

John Tate, who ran CFL until taking a leave of absence to manage Paul's 2012 campaign, remembers walking offstage at the rally. "I'll never forget this. I spoke for literally 30 seconds to a minute, and for the first time in my life walking off that stage, there were people standing there wanting to shake my hand and get autographs," he says. "Afterward I called my old friend Tom Lizardo [a former chief of staff for Paul's congressional office] and said, 'Wow, this is weird.' And he said, 'You really have to understand that this is literally to a lot of people a new American revolution.'"

Rye was filming Paul's speech to the arena-filling crowd of more than ten thousand, all there for a counterprogrammed political convention for a failed primary challenger—when does this sort of thing *happen* in America?—and thinks he saw a man's dream coming true. "Not just for the campaign, but a thirty-year culmination of what that guy stood for. This guy put everything he had, with people laughing at him and mocking him, and I just thought, this is going to live in history. You don't see this kind of thing that often. A guy fights his whole life for something, beat down along the way, and finally gets people to recognize what he's been saying. It was one of the best speeches I ever saw him give—this guy, seventy-three years old with this stadium full of young people who came to see him, very much a rock star, but talking about ideas. You could see he realized he's finally getting his due—though he doesn't like to think like that, he's clearly half-embarrassed by the cheers at time.

"Nearly an hour, no prompters, he's just jamming. He knows how to play, and he's up there jamming. And when he's done, he just bends down and shakes people's hands for five minutes while the entire arena goes crazy. And as he's leaving the stage, he has this look of complete humility, waving good-bye and so thankful that you people came, that you actually care about what I have to say. You knew this dude is a human being, no fucking political robot."

I ran this interpretation by Paul in 2011; did he remember

this moment as extraordinarily special? That's not the kind of guy Ron Paul is. He's just continuing to carry on the mission he feels his supporters have in some sense laid on him: to keep talking about liberty.

In the end, Paul went into the 2008 Republican convention in Minneapolis with somewhere between twenty and thirty-five pledged delegates, third to McCain and Huckabee. Paul refused to endorse McCain and so was not invited to appear. Paul tells me they told him he could show up if he wanted and some nice gentlemen from the Republican National Committee would give him his credentials, walk him around the floor, and walk him out. No thanks. Paul delegates had their Paul schwag taken from them before they could carry it onto the convention floor. Members of the Massachusetts delegation who intended to vote for Ron Paul were pressured in an atmosphere of high security and paranoia to switch to McCain, with an implied promise of possible Paulite influence within the state party later only if they went along. Many recall what should have been delegate votes for Paul from the floor not being announced or recorded. It's hard to say precisely how many delegate votes he actually had, and by the official rules, since he didn't have a plurality of delegates in at least five states, he wasn't technically even eligible to be nominated by the convention, but, well, it was nowhere near enough.

Why didn't the campaign do better? A great many of the activists who had recently boarded the Paul boat thought this question demanded an answer. To those of us who had been watching the Ron Paul story unfold for decades, the more important question was, How did it do as well as it did? As campaign fundraiser Bydlak asks, "What other libertarian candidate has ever managed to raise close to thirty-five million dollars and be somewhat competitive in many primaries?"

Anthony Gregory, an anarchist libertarian analyst from the Independent Institute who is very popular with the LewRockwell .com crowd, was talking to Paul about a possible job with Campaign for Liberty (he ended up writing content for their website for a few years, but doesn't anymore). He remembers telling Paul that he wanted to write him in on his California ballot but he knew Paul had been telling people not to do that, so . . . Paul just told him to do what he wanted: "I don't like to tell people what to do."

Lots of people wrote Paul in, in California—17,006 to be exact. Despite not being the Republican nominee, Ron Paul was on the ballot in two states for president. Trent Hill in Louisiana didn't feel like giving up after the exhilaration of Revolution Season in 2007–2008. He found out you didn't need the candidate's explicit permission to get him on the ballot, and did the paperwork rigmarole to create an impromptu "Louisiana Taxpayers Party" with Ron Paul as their presidential candidate and Barry Goldwater Jr. for veep. A breakaway faction of the Constitution Party in Montana also took off the national party's nominee Chuck Baldwin and stuck in Ron (even though Ron had endorsed Baldwin, after briefly merely advising his fans to vote for any third party candidate). Between the two states, Paul racked up nearly twenty thousand ballot votes.

Could Paul have done better? His old congressional aide J. Bradley Jansen, who walked away from his work on delegate wrangling with the 2008 campaign angry at what he thought was mass incompetence on the part of an operation he is convinced was never prepared to seriously compete, is confident Paul could have fought such a good fight in the second half of the campaign, after everyone but McCain had disappeared, that Paul would have been guaranteed control over the veep slot, if only the campaign had spent every cent it had rather than passing $3.5 million on to Paul's congressional campaign committee. Another old Paul congressional hand, Bruce Bartlett, thinks Paul could have done better focusing on his uniquely anti-Republican positions, such as being against the Iraq War and the drug war, to the exclusion of all that money and Fed stuff. "But every time he has a forum, he's got to talk about abolishing the Fed, which made him sound like a kook. But money really *is* the love of his life—and not in the base sense."

Many libertarians otherwise fond of Paul echo the complaint that "he went crazy with the Fed stuff." I was told by an old hand at a Washington, D.C., think tank that his younger colleagues, inclined in some ways to like Paul, thought that his obsessions with inflation and central banking were politically tone-deaf and economically mistaken. But the tremendous objective success of the "End the Fed" meme, Paul's rise to chairmanship of the Financial Services Subcommittee on Domestic Monetary Policy in 2010, and the general realization in both punditry and the economic profession that the Fed has been at best feckless and at worst destructive in both the lead-up to and aftermath of the economic crisis, has made that part of Paul's message the most seemingly prescient, sensible, and necessary.

Something Paul said to me near the beginning of the campaign, in October 2007, was still true, and indeed even more so, at its end: "Something very strange is going on and I don't think anyone has fully comprehended how big it really is, and each day it's bigger than I ever thought it could be."

★

CHAPTER SEVEN

ONE, TWO, MANY PAULS

The Ron Paul Revolution continued to take every field it could occupy in political culture while Paul returned to Congress, rebuffing suggestions he might want to do a 2010 Senate race.

The Campaign for Liberty managed to raise three times the money in 2009 it did in 2008 and racked up hundreds of thousands of signed-up activists; Students for Ron Paul morphed into Young Americans for Liberty and began organizing student groups and running boot camps for liberty activists in the basics of real-world campaigning. The "End the Fed" meme went from a chant at Ron Paul rallies to its own mass movement, launched in November 2008 by Los Angeles-based Paul grassroots activist Steven Vincent, and for the first time Paul's perennial "Audit the Fed" bill actually passed the House in 2010, though not the Senate. Campaign for Liberty and the grass roots did execute a grassroots pressure campaign to get it through the House, and it actually worked. Every single Republican in the House voted for an audit of the Fed, and Ron's book on how the Fed caused economic booms and busts with its profligate interest rate and monetary policies, *End the Fed*, hit

the bestseller list, a follow-up to Paul's campaign-year number
one, *The Revolution.*

John Tate, CFL's chairman, credits CFL for the unprecedented
success of the "audit the Fed" bill. Conservative movement activists from the Tea Party world, not directly connected to the Paul
machine, have also told me that the bill's success demonstrates
CFL's clout. "We started with an issue where 75 percent didn't
know what the Fed was," Tate says, "and now 75 percent want to
audit it. That is one way to look at CFL and realize we can claim
without any hesitation it has been a huge success." Tate's comment reminded me of one of my favorite things I've heard from
any Paul grassroots activist: Des Moines–area Paul supporter Allen Huffman was talking about the strange aspect of opening
your eyes to a whole new world, a different world from the one
most of your friends and colleagues seem to live in, that comes
with embracing Ron Paul. "Before Paul, I didn't know what the
Federal Reserve was." Huffman paused, and thought about how
much more sinister and off-balance the world seems now that he
knows about the Fed's power to harm the economy: "I wish I still
didn't know what the Federal Reserve was!"

Every Ron Paul fan ends up knowing about the Federal Reserve. Paul's beef with the central bank is a by-product of his
longstanding interest in the works of Mises and Hayek. Paul, like
the economists he admires, thinks it a mistake to have a giant
government-run institution trying to fix prices—in this case, interest rates, or the price of loaned money, which is the Fed's main
mechanism for pursuing its stated goals of economic growth,
high employment, and relatively stable prices. As a critic of state
power, Paul also worries that once a government has total control
over paper money that it can create at will, it becomes too easy
and too tempting for the state to *spend* at will. Cash unbacked
by gold will flow to help the government out of its jams, pay for
its wars, and appease its most powerful private constituents. To

Paul, this danger makes the Federal Reserve, central banking, and "fiat" money *the* key libertarian issue. If the government can manufacture all the money it wants, the fight for limited government is over before it begins.

Central to this critique is the Austrian business cycle theory, which helped win Hayek his Nobel Prize for economics in 1974. Hayek argued, to simplify a very complicated line of thinking, that low interest rates set by the Fed fool investors and builders into thinking that consumer demand for future goods is higher than it actually is. Cheap money makes producers more likely to launch long-term projects and take on long-term expenses. When low rates are a product of government intervention, rather than a market expression of people's desire for long-term goods as reflected in their willingness to save now in order to consume more later, those long-term projects—for example, building and buying homes—will turn out to be unsustainable "malinvestments." Prices in those areas will plunge. Everyone will start to realize that resources were funneled to unprofitable ends. An exaggerated boom will turn into a catastrophic bust.

Austrians believe increases in the money supply don't always manifest in economy-wide rises in the consumer price index, the standard definition of inflation. The excess cash might instead flow into specific areas of the economy, depending on real-world factors that vary from case to case. In the housing boom and bust, those factors included mortgage lending standards, the actions of the government-created mortgage holders Fannie Mae and Freddie Mac, and reckless securitization of mortgages. In the Fed skeptics' story about the last decade's economic expansions and contractions, the housing bubble was a deliberate effort by the Fed to stave off economic troubles that began when the tech-stock bubble burst in 2000. And no one has made the skeptics' story more popular than Ron Paul and his organizations.

"Ron wanted Campaign for Liberty to be just educational,"

Jesse Benton says. "But we drug him kicking and screaming to make it a grassroots lobbying group, and one that trained activists—we've trained hundreds of activists in every state, giving them the tools to win legislative fights and electoral fights at the local level and state level." CFL, while an outgrowth of the Paul movement, has grown beyond it, Benton says, with nearly 100,000 donors who had not previously donated to Paul's own campaigns. "People don't give to CFL just because they like Ron Paul," he says. "They give to it because they like our issues."

Paul's chief of staff, Jeff Deist, sees jealousy on the part of some colleagues over Ron's new national prominence. "It really bugs them when someone comes up to them and asks them to be more like Ron Paul. I get other chiefs of staff handing me books [from their constituents] for Ron to sign all the time." The presidential campaign convinced his Republican colleagues that Paul represents a bigger wing of the party than they thought— "and now that Bush is gone, criticism of the status quo is easier" among Republicans, Deist says, since "it is now directed at Obama." More than a dozen colleagues are regular guests at informal luncheon talks that Paul hosts, in which he has libertarian scholars speak on their areas of expertise; guests have included the Cato Institute's David Boaz, libertarian journalist Jim Bovard, tax expert Dan Mitchell, terrorism expert Robert Pape, and Austrian economist popularizer and historian Tom Woods. Michele Bachmann used to show up regularly. One Paul staffer suspects Bachmann was strategically trying to glom onto the Paul movement, prior to her own presidential run and her public claims to read von Mises on the beach.

The political and media culture changed after Paul's 2008 run. Financial bestsellers by the likes of Woods (who only wrote the book after he suggested Ron write it himself) with his *Meltdown*, which explained the Fed's culpability in the financial and housing crisis, and Charles Goyette (a former right-wing talk show host from Phoenix who went too Paulista on the subject of the wars in the Middle East and lost his show) with his *Dollar Meltdown*, spread a very Paul message without being saddled with whatever negatives one might have toward the candidate himself.

Woods wrote one of those *New York Times* bestsellers that the *New York Times* prefers to ignore, and he has since become the biggest speaking-circuit superstar in the Paul world who isn't named Paul. He can issue YouTube videos of himself sitting at a desk explaining some abstruse economic or historical point for 27 minutes and get tens of thousands of views. (The ones where he is explicitly defending Paul from some ignorant or unjust attack get the most views.) One of the more noticeable changes in media culture was the rise of Judge Andrew P. Napolitano, a former judge with the Superior Court of New Jersey and senior judicial analyst and anchor of *Freedom Watch* on the Fox Business Channel.

Napolitano came by his hyperconstitutionalist ideology honestly, as a state judge in New Jersey trying criminal cases and "observing the government at close hand and concluding inexorably that most of the time they were evading and avoiding the Constitution rather than honoring it. I saw the utter scandal of prosecutors bending and cheating and police outright lying in an effort to get around the Constitution." He saw a great deal of this sleazy behavior in the context of the war on drugs. Napolitano already had libertarian and constitutionalist tendencies; when he was a young man in the early 1960s, "I was drawn to Goldwater the way young people today are drawn to Ron Paul, thinking he's a godsend, such an antidote to monster government. Though the monster government of 1963 seems quaint and constitutional by today's standards."

Napolitano began doing legal commentary on Fox News in 1998, during the Clinton impeachment. He became a regular guest host on various other Fox shows, including Glenn Beck's.

Then, in 2010, he was given his own nightly hour on the companion channel Fox Business News. Napolitano was already known as the reliable guy on Fox News who went beyond mere Republican propaganda to, as Napolitano explains his role, "reliably explain the libertarian, free-market, individual vs. government side of things" as well as "the basic Econ 101, Hazlitt, *Economics in One Lesson*, those first principles of economics." The Paul fans in the Fox audience already had their eyes on him, seeing him as the good guy on the Fox team. So when Fox refused to allow Paul into the crucial pre–New Hampshire primary debate, "there was a tremendous outcry, and Fox received well over 175,000 emails complaining, and unfortunately every one of those was cc'd to me."

Perhaps, Napolitano says, "that may actually have caused [Fox News president Roger] Ailes, who is a marketing genius, to say, if they feel so strongly about Ron Paul, why not give them what they want? Unleash the judge and bring those people back. And a lot have come back, and not just to watch me because they like what I say. They come back to watch a lot of Fox products, and you know what? Suddenly Ron Paul is not so fringe anymore. Audit the Fed, end the Fed, people take changing our foreign policy seriously. There has been tremendous progression since 2007."

Napolitano recalls a scene from New Hampshire in the first Paul run, where Sean Hannity was running for cover to avoid a mob of angry Paul fans. "Now Ron is Sean's guest and it's as warm and amicable as can be. Ron had left the fringe world and entered the mainstream here at Fox."

Napolitano keeps his eye on positive developments in Congress and says he knows of up to fifteen congressmen whose voting records mark them pretty much as hard-core faithful libertarians on a range of issues from the Patriot Act to the debt ceiling to the war on Libya. And nearly double that number at least seem to lean libertarian, if not with steadfast reliability. "And I don't know if

any of them would be there if not for the persistence and education Ron has caused to come about through his personal, persistent, and continual education of the public and other members of
Congress."

In 2008 and 2010, dozens of self-styled "Ron Paul Republicans" sought office with the Republican Party. A New Hampshire Paul campaign volunteer named Jim Forsythe (also an Air
Force veteran who had participated in no-fly-zone enforcement in
Iraq between the two Iraq wars) became a state senator in New
Hampshire, and a handful of self-conscious Paul devotees are
also in the New Hampshire statehouse, where they've fully le-
galized knives, loosened homeschooling regulations, and helped
actually cut overall spending. Forsythe points out that Paulian
candidates have to hone their pitch to specifically local needs and
concerns: "you can't just win state office by saying you are for
liberty." Four Paulians won GOP House primaries in Maryland
in 2010, though they were all defeated by Democrats. Paulite B.
J. Lawson was the GOP candidate for a South Carolina House
seat in 2008 and 2010, though he lost both times. A huge Paul
supporter, John Dennis of San Francisco, was the Republican opponent to House Speaker Nancy Pelosi in 2010, though he failed
to unseat her. Paul fan and investment world bad boy Peter Schiff,
who was also predicting a housing bubble collapse from Paul's
Austrian perspective, made a high-profile run for the Republican
Senate nomination from Connecticut in 2010, losing to wrestling
mogul Linda McMahon, though Schiff pulled over 20 percent
of the vote in a three-man race. A few candidates supported by
Paul's LibertyPAC won state House seats in Iowa, including Kim
Pearson, who publicly announced her intention, in September
2011, to recruit primary challengers for fellow Republicans she
thought insufficiently conservative about things like gun rights.
The willingness and ability to field candidates on the part of the
Paul revolutionaries is clearly there; the ability to win seats, not

so much yet. While it was too early to be sure as this book was being written, various people close to Paul say there should be even more Paul-identified and Paul-supported candidates making primary challenges to Republicans in 2012 for federal office. One already on the horizon is Mike Doherty, a state senator from New Jersey contemplating a run for a federal Senate seat in 2012, who people in Paul's camp think could become a national cause célèbre for Republicans interested in small-government red meat. Benton says Paul likes to know any candidate seeking his endorsement could raise at least $50,000 on their own.

The Paul Revolution has one clear federal victory already: Michigan representative Justin Amash, a guy who votes no even where Dr. No votes yes at times, and explains his every vote to his constituents on Facebook. A true child of the revolution, in spirit and technique, Amash became a Hayek and Bastiat fan while an economics student. He became disgusted with the sameness of the Republicans and the Democrats, won a state house seat in 2008, and instantly decided to try for federal office.

He was a fan of Ron Paul from 2008, and sought his endorsement for his 2010 federal House run. He went to visit Paul in Texas, and "he wanted to know I thought I could win it without him. I think he doesn't like it when people come and ask for his help who think he is going to carry them to victory just because they are big fans of his and they run as Ron Paul Republicans. He told me when people come in saying they are Ron Paul Republicans and want his endorsement, he tells them, 'whatever percentage of the primary vote I got in your district, that's how much I can help you.'" And that's usually not that much.

Amash has been a bit surprised at how little his colleagues seem to know about what they are doing: he cites votes of more than four hundred lawmakers where he can't find any fellow House member who can even explain what the bill they all just voted for will even do. He's working on a balanced budget amendment and is doubtful the House will ever get its fiscal house in order without one. He finds that when his constituents are mad at him about anything, it's usually a generic anger at their vision of a Republican. When he explains his own votes, he says, they are on his side. He was redistricted this year so will be facing a new constituency in 2012, but he's confident he can keep his seat. Amash doesn't want to speculate about his possible future role as the national political leader of the liberty movement now surrounding Paul until Ron Paul is no longer in office, or running for it.

The Paul Revolution's relationship with the Republican Party is still fractious. From local meetings to the national convention in 2008, plenty of Ron Paul activists have tales of disrespect or even abuse from GOP regulars—from cops being called at meetings overwhelmed by Ron Paulers in Missouri to parliamentary rules being abused to hurt Paulites' chances in Louisiana to a general sense of get-out-of-our-face disrespect everywhere from Texas to Oregon.

Jeff Greenspan, who has been a Republican county chair in Arizona and worked the caucus process for Paul in Nevada and elsewhere, says that any new forces making an incursion into an existing realm, especially with the attitude that they are out to change it, should expect resistance, and just be prepared to put in the hours it will take to overcome that resistance.

As Florida Republican activist and Paul fan Phil Blumel observes, "In 2008 when the Ron Paul folks first showed up they didn't know how to deal with people professionally, and in the last few years that's gotten better. Some of that is just them growing up. In 2008 the Ron Paul people were all really young. Now in Florida the Paul contingent is respected in a way it was not in 2008, and they dominate in some counties."

The people who fill these party positions, though the average voter will never know who they are, "have a lot of influence on who wins primaries," Blumel says, "because the movers and shakers of the party are listened to by lots of the rank and file. They will follow their votes. So when you have Ron Paul presented as a respected alternative by normal people like them working in their party like them, it makes it more possible for other Republicans to publicly support him. Paul people were being laughed at and barred from things in '08. The chair of the state party saw them as a barbarian threat." That chair has since been indicted for embezzlement, and being a Paulite is no longer such a liability in Florida. It doesn't mean that Paul will win the state in 2012—there are too many mainstream Republicans who "despise and distrust all dissident candidates, Bachmann or Paul or whoever else," but where it used to be derisive laughter aimed at Paul and his fans, "now it's at least polite respect even if they'd never vote for Ron Paul."

I hear similar stories from Paulites across the country. Chris Rye is treasurer of his county party in Wisconsin, and the chair, vice chair, and secretary are all Paul fellow travelers as well. In Rye's experience, for some of these positions, pretty much being the people who show up guarantees you can win local party office. The Pat Robertson troops legendarily did this after his 1988 run and made the Christian right as influential within the GOP as it is. As Rye sees it, liberty-minded would-be Republicans shouldn't be concerned about obeisance to Paul per se but about moving into the party and "trying to root out the entrenched guys that just wanna win that delegate seat for their own ego, who don't really care [about policy] as long as a Republican wins. That mentality has infected everything." The country needs a GOP of "troublemakers; the go-along-to-get-along hasn't done the country any good. We need to replace those people with passionate activists from the grass roots."

Rye doesn't personally behave like a hostile firebrand to his non-Paul-following GOP colleagues, and is finding that people who were pro-McCain in 2008 are coming around, especially on foreign policy. The one blessing of President Obama for the right-wing antiwar movement: it has made it okay for Republicans to be for peace. And as Paul people around the country have found, in an aging political party their merely being young and having the enthusiasm and energy to actually do the work that needs to be done to keep a party functional can go a long way to over-come ideological prejudice (though not necessarily all the way to overcoming it). Rye still remembers with some bitterness how at a Wisconsin GOP convention a plank that a Paulista introduced about hemp legalization was openly mocked onstage by the chair (while Paulites put a Ron Paul T-shirt on a McCain cutout and took pictures).

Blumel puts it bluntly: "By definition, for every libertarian who joins the Republican Party, that makes the Republican Party more libertarian." Rye agrees: a political party is made up of the people who show up, and he wants Paul people to show up for the Republican Party. Iowa is already a very Ron Paul state, with various county committees filled with Paulistas.

Paul's 2012 campaign manager John Tate, whose background was in the right-wing–identified Leadership Institute and the Right-to-Work Committee, sees the growth in Paul influence in the Party. "In '07–'08, we couldn't get any local Party people to invite Ron to speak; now we get literally hundreds of such invitations a week, and part of that is [because] his message is more popular now," Tate says. "But it's also that a lot of our people are now running those local Party machines and they get to decide who comes in and talks to them. Local parties can also help control the flow of money to candidates. It varies state by state and district by district but local parties can give and do a lot for candidates."

Although Paulistas across the country are sanguine about their future in the party, longtime hard-right fundraiser Richard Viguerie, who was helping Paul in the mid-1970s, is dubious that the Paulistas will gain much traction. He thinks they just aren't

community-oriented enough, aren't the types who have the positions in business organizations or church groups or Lions Clubs
that give them the connections fellow Republicans will respect.
Just not good enough joiners in general. But then he vacillates a
bit, remembers the old days, and wonders: "When we were trying
to get Reagan to beat the Ford/Rockefeller wing in '76, we realized we needed a new leadership unfettered by old ties and old
relationships. And the Ron Paul people are unfettered by old ties
and old relationships. When they come in they are likely to make
common cause with the Tea Party and make things uncomfortable for Republican leaders, the Boehners and McConnells, and
we can't go after them until the party is more occupied by people
who are unfettered."

Viguerie mentioned the Tea Party. You can't study Ron Paul's
2007–2008 revolution without noticing its reliance on the idea of
the new Tea Party. His historically unprecedented one-day money
draw of over $6 million happened on the 2007 anniversary of the
(original) Tea Party. That same day, huge Ron Rallies were held
in Boston itself, across Texas, in Santa Monica—even Hawaii and
France. Ron himself showed up at one in his home district and
tossed a barrel marked "Iraq War" into the water.

Ron's son Rand Paul led the festivities in Boston's Faneuil
Hall, with a thousand people daring a foul winter blizzard, and
told them how *they*—all Ron Paul fans know who *they* are—
called them the Ron Paul rabble, and laughed at them. "And they
are not laughing now."

Since that big Tea Party day in 2007, a mass political move-
ment has arisen, one that is transpartisan and dedicated to a very
Ron Paul message of out-of-control government, no bailouts, and
changing the game in Washington. Ron's son, currently Senator
Rand Paul, Republican from Kentucky and head of the Senate's
Tea Party caucus, was the star attraction at the 2007 Ron Paul
Tea Party rally.

The Paul movement has been one of the political phenomena of the modern era least noted by mainstream media and political scientists. The Tea Party has been one of the most noted. Entire books have been written on the modern Tea Party phenomenon that don't see fit to mention that the first one was a Ron Paul movement event. Yes, the first modern strongly antistate, antitax, and antispending movement of the modern era was Ron Paul's, and Paul's son Rand won his Senate seat by explicitly connecting himself to the Tea Party movement, wearing its mantle and fighting as its champion. The Paul campaign itself has taken to calling their man "the intellectual godfather of the Tea Party."

How to link and entwine the Paul movement and the Tea Party movement—indeed, what makes the Tea Party folk not embrace Ron as their candidate—is a tricky question, one that even most people involved wouldn't know how to answer. Well, Ron Paul is just . . . Ron Paul. Over there in his own world. The Tea Party is the Tea Party, in theirs—even though Rand Paul wrote *The Tea Party Goes to Washington*. In every Ron Paul event I attended in the fall 2011 campaign season, from Los Angeles to Nevada to Iowa to New Hampshire, I saw almost no one self-identifying openly as a Tea Partyer.

Dan McCarthy, who worked in 2008 on the Paul campaign's media staff, and now edits the *American Conservative* magazine, recalls that "in 2008 after the official campaign had ended, there were lots of activists on Facebook, wondering what the next thing to do would be. Campaign for Liberty was not quite off the ground. There were lots of people saying, 'Let's have a big tea party this year on the anniversary of the Ron Paul 2007 money-bomb.' So the idea of a Tea Party was in quite general circulation among libertarian activist types, libertarians reawakened by Ron Paul, who wanted to keep things going, find new symbols and new ways to fan the flames. The Tea Party phrase and idea was circulating a lot on Facebook.

"Maybe Santelli [Rick Santelli, the CNBC announcer popularly credited with inventing the modern movement in 2009] said it on his own and hadn't heard these other rumblings. Maybe it was a series of coincidences. But there is a good claim historically as well as on the fact that Ron Paul was already talking about Tea Party small-government issues in '07, for calling Paul the father of the movement."

It's likely true that very few of the people flocking to Tea Party events in the wake of Santelli were doing so with any conscious knowledge of or influence from the Ron Paul Revolution. Historians are good at tracing clear lines of influence, or precursors, to any given phenomenon, but intellectual historians sometimes forget that most people don't know intellectual history.

The Tea Party as it has evolved has some Paulian aspects, but also lacks some important parts of the Paul message. Paul himself has upbraided the Tea Party masses for not understanding that smaller, affordable, constitutional government requires ending overseas empire even more than it requires cuts in entitlements.

An Oregon-based Paul activist, Mark Hutto, has watched the rise of the movement, and mocks the Tea Party type who says, "I want small government but I want to continue the war on drugs. I want small government but I'm worried about what others are doing in their bedrooms. I want to have marines in every country in the world but I want small government. I want to control every facet of your life if you are doing something I don't enjoy doing, but I want small government. I don't want cuts in Medicaid or Social Security but I want small government. I'm going to vote for Sarah Palin but I want small government."

McCarthy notes that Tea Partyers and Paulians differ in tone and attitude as well as in their full package of political beliefs: "I saw little of that ticked-offness" so common among Tea Party activists in the Paul movement. "Ron does not like nasty or hostile politics," McCarthy says. That was "a restraint" in their media operation during the 2008 run. "We were not allowed to attack other candidates, and even mild stuff like press releases indicating another candidate's hypocrisy or difference between their words and actions, we could do very little of that. At one point we did do a release more sharply worded than usual and Ron got ticked off; there was some sturm und drang about it. He doesn't like that nasty edge in politics. He wanted to keep everything as highly toned as possible. So some of the attitudes of the Tea Party worry him." (By the 2012 campaign, Paul did give his campaign operatives their head in making ads that specifically targeted his opponents.)

Even in their angry populism and their concern with economic issues more than foreign policy ones, McCarthy thinks the Tea Party might be reachable through "arguments on the diseconomies of war." The Tea Party types might not be Ron Paul anti-empire activists, but they are probably amenable to "jettison[ing] a little militarism" for the sake of a balanced budget.

"Ron wants to be friendly with the Tea Party," says his congressional legislative director Norm Singleton. "But he doesn't like the idea of a congressional Tea Party caucus. He thinks what's good about the Tea Party is that it wasn't a top-down D.C. movement, it was bottom-up, and it should not be co-opted" by anyone in Washington, even him.

Rand Paul has already "co-opted" it in a sense, claiming identity with the Tea Party movement in his book telling the story of his unlikely political success in 2010, *The Tea Party Goes to Washington.*

Rand Paul was not supposed to be senator. When Jim Bunning, then recently famous for trying to hold up unemployment benefit extensions, decided to throw in the towel and retire at the end of his term, the elder Kentucky solon Senate Minority Leader Mitch McConnell and the rest of the GOP establishment thought the seat should go to Kentucky secretary of state Trey

Grayson. Grayson was the anointed; Rand Paul was an obscure eye doctor with the weird dad. But between late summer and December 2009, the polling for the 2010 Senate race in Kentucky whipsawed from Grayson beating Paul by 15 percent to Paul beating Grayson by 19 percent.

How did he do it? Rand was able to only slightly edge out Grayson in money raised, by tapping into Ron's national audience and moneybomb technique. Grayson's team tried the libertarian baiting, with a public statement saying that "maybe Ron Paul's skills as a career politician will help his son Rand explain to Kentuckians how closing Guantanamo and releasing the prisoners will make us safer and how a pro-choice marijuana advocate will best represent Kentucky Republicans as their Senate nominee. . . . The truth is that Ron and Rand Paul are not conservatives on national security and social issues and are completely out of touch with Kentucky."

Well, then! That tactic didn't work. Rand didn't run on those more outré ends of the Paul message. He ran against the bailouts, stressing Tea Party disgust with two-party business as usual. He believes in the Austrian business cycle and ending the Federal Reserve but hasn't often led with those issues. He wouldn't have voted for the Iraq War if he had a chance, but he doesn't complain about empire much, and he alarmed some of his dad's anti-neocon fans when he deigned to meet with neocon leader William Kristol.

In the campaign Rand Paul was not lacking in national support—more insurgent-oriented groups such as FreedomWorks, Concerned Women for America, and good ol' Gun Owners of America were for Rand. Erick Erickson, head of the red-meat right-wing blog RedState, which banned Paulites in the 2008 campaign, said he's "one million percent on board with Rand Paul."

But as Rand's campaign manager for the primary, David Adams, told me, in both money and media the campaign at the start was unquestionably built on the story of the son of Ron Paul; the impact Ron Paul had made nationally softened things up for Rand in unquantifiable ways. The Tea Party was the vehicle that dragged him over the finish line; every event, Adams recalls, drew hundreds more people than they anticipated. Two thousand ten was the right year for this antiestablishment campaign to win. Rand recalls that at the beginning, he couldn't even get invited to local Republican Party meetings; by primary time he beat the establishment hero Grayson by 24 points. (He later beat the Democrat by 12.)

Right after winning the primary, Rand Paul stepped into a ready-made libertarian trap. Rachel Maddow got him to get all libertarian-philosophical on the issue of the 1964 Civil Rights Act, and Rand mused about how he didn't think forcing private businesses to associate with people they didn't want to was necessarily government's responsibility. After a predictable media furor about how outrageous this position was, Rand later said that he considered the act on whole a good thing and certainly had no intention of trying to overturn it—which should have been obvious to anyone not out to get him in the first place. He learned that media is not his friend and that he needed to be less freewheeling in public statements. He retreated from the press for a while to regroup, even backing out from a planned appearance on the high-profile NBC show *Meet the Press*.

He's gotten smoother at being Rand Paul. As Jim Antle notes in a *Reason* magazine profile, Rand Paul has a gift of emphasis that allows him to say things similar to his father on foreign policy but have it feel different, leading into a question about how terror attacks on America might be blowback from our own past mistakes with "the most important thing to say from the beginning is that if someone murders your family, it is their fault. . . . We say these people attacked us, it is their fault. . . . We are not to blame for people attacking us." Rand told Antle that "I don't agree with [Ron Paul] all of the time" and that he's trying to build and represent a small-government coalition that might be bigger than just Ron Paul fans: "I am not trying to splinter off into a smaller and smaller group. I am trying to create a force that can win an election."

Frankly, Rand's better success with the normal Republican/ cosmopolitan libertarian audience confuses me—I have no clear explanation for it except that perhaps Ron's presence on the national scene for so many years has had enough peculiarities that anyone not with him 100 percent has gotten annoyed with him somewhere along the line. Rand is even as anti-abortion as his dad, and less sharp on civil liberties, having approved of keeping Guantanamo open (and in one awful gaffe on Sean Hannity's show seemed to suggest he felt that attending subversive rallies was sufficient reason to be arrested—in fact he was merely discussing using intelligent signs of probable cause before unleashing government investigative efforts on everyone). But Sarah Palin endorsed him, and mainstream libertarian groups such as the Cato Institute seem to adore him unreservedly, without the hedging that always surrounds Ron Paul. Meanwhile, that Internet center of Ron Paul love, LewRockwell.com, barely mentions Rand. Rand seems to be straddling a reputational divide, mostly keeping the Paul masses on his side while appealing to people his father has so far failed to win over.

The narrative has become that Rand has . . . something—a smoothness, an ability to frame himself—that his father lacks. A Senate race is harder to win than a House one, though Ron likes to joke that he tells his boy, Congratulations, now, if you really do well with this job maybe someday you can make it in the House of Representatives! An NBC reporter once asked Ron Paul at an Iowa speech if his son's political success had taught him anything. Paul answered that he will just keep doing what he's always done—tell people the truth. They will respond. Ron Paul does not think he has anything to learn about politics.

Once in office, Rand delighted right-wingers and constitutionalists who were dreaming of a fresh face to make their case for them, balls to the wall, in the media. One right-leaning journalist who didn't want to be quoted being so gushy about a politician told me he feels like he's practically in love with Rand Paul, thinks he's the greatest thing ever on the American scene. Rand's very first speech as a senator was about how compromise is not always the highest value. Rand will tell Anderson Cooper that it's more dangerous to the country's faith and credit to keep adding more debt than it is to face up to our troubles with the "temporary inconvenience" of *not* raising the debt ceiling.

Rand holds Senate hearings to highlight the "jackbooted thug" side of government regulatory enforcement, how people's lives get ruined over dampness on their land that makes it a "wetland" or for selling rabbits without permission, or for using a certain type of wood in their guitar factories. Rand will try to get the Senate to vote on Obama the campaigner's declaration that the president does not have the power to unilaterally declare war, which would take away Obama the president's power to keep illegally fighting the war in Libya he started unilaterally. He'll be the only guy to actually hold up Patriot Act renewal to force a vote on an amendment to protect gun records privacy while being attacked with fearmongering that the republic will fall if the act isn't technically in effect for even two hours. Didn't the Fourth Amendment do the job for us for centuries? Rand wonders.

So far Rand has found that the power of one senator to effect change is small, of course, but with comrades like Mike Lee of Colorado (also a recipient of Paul LibertyPAC money, though he says he came to his constitutionalist conservatism independent of a direct Paul influence) and Jim DeMint of South Carolina on his side, he's actually gotten his proposal that the budget be balanced in five years up to a vote (it was crushed, 90–7) and is working on making sure there are at least public debates on important controversial issues like the Patriot Act and No Child Left Behind. As thrilling as it was for the revolution to get its own senator, the main goal is bringing more voices of liberty into the political show—it's not enough by itself to effect real change. As Ron Paul always thought. That has to come with the changing of many, many minds—of the people who pick the politicians.

─────── ★ ───────

CHAPTER EIGHT

AMONG THE PAULISTAS

R on Paul people are not exactly a different breed, but they are a distinct one. Their variety in socioeconomic status, original political outlook before discovering Ron, and choices of how to move through the world has been much noted. Students, housewives, street anarchists, soldiers, small-business people, even government employees have all found themselves dedicating their lives, fortune, and sacred honor to pushing the message of Ron Paul.

They tend to forge their own world of Ron Paul information, getting news and news recommendations from the *Daily Paul* and Ron Paul Forums and LewRockwell.com and from each other's Facebook feeds and blogs. There is a heavy sense that the "mainstream media" isn't going to give them the news and analysis they value or trust; this perception is fueled by, among other things, the media's treatment of their man Ron and their ideas. It is widely believed, as Jon Stewart made comic hay from, that the media are more or less deliberately blacking out Ron Paul. His $3 million moneybomb (spread over three days, not the usual one day) was built around this idea in October 2011; Paulistas began turning the name Ron Paul into an unreadable black bar on their own sites as an ironic comment.

The Ron Paul Revolution is about more than Ron Paul—as Ron will himself insist both explicitly and in his use of the words *us* and *we* and *our ideas* in public. It's also about the people who want you to know about Ron Paul. I spent some time seeking out Paul people in places where they gathered just as people, rather than just to hear or show support for Ron Paul.

Ron Paul fans are everywhere. They are inclined to try to be prepared for everything and anything. Far from utopian—an accusation often tossed at those who believe our country should have far less government—Paulites trend more toward the *dystopian*, nurturing a dark Paul-buttressed vision of a "normal" world far more fragile, far more ready to break apart in an ugly mess, than most typical Americans recognize.

Overspending, overborrowing, overintervening, overinflating: in Paul's vision, all these things mean a looming crisis, one so severe and disconcerting that the debt ceiling mess in the summer of 2011 was merely a sanitized preview that glided discreetly over the parts too intense and harrowing for general audiences. As longtime Republican hand Drew Ivers, campaign chairman for Ron Paul in Iowa, had a habit of telling the crowds at Ron Paul gatherings, the crisis was not something we should fear will fall upon our children or grandchildren; we are in it *now*.

It thus made perfect sense that I would find a concentration of Ron Paul fans and interested parties in early June at a festival on the water called Ephemerisle, dedicated to figuring out how to escape the travails and oppression of modern civilization through particularly colorful means. This was the third such gathering. It began in 2009 as a sort of proof-of-concept experiment sponsored by the Seasteading Institute, a nonprofit organization launched by Patri Friedman, the grandson of libertarian economist hero Milton Friedman, and son of the more obscure economist and political philosopher David Friedman, who is nonetheless popular in libertarian circles. David wrote about the possibilities of a working, happy society with no government at all, in the sense we understand it.

The Seasteading Institute's biggest seed money came from high-tech and finance visionary Peter Thiel, a PayPal founder with a yen for science fictional visionary charity. Thiel has also sunk some of his huge piles of cash into causes such as hyperlongevity research and space travel. Why the Seasteading Institute? To promote the construction and occupation of new floating nations in international waters, of course.

And why would people dedicate themselves to this curious cause? For reasons that Ron Paul people understand. Patri Friedman thinks the solution to the sclerosis and bad policy outcomes of modern government that his grandfather and father inveighed against is the same as the optimal solution to sclerosis and bad outcomes in markets: more competition.

How do you compete with governments that claim control of every inch of the planet fit for humans to live on? You make new places for humans to live on—artificial floating platforms in the open, unclaimed ocean. The Seasteaders plot and plan businesses that might find a special comparative advantage in operating outside the reach of governments' rules; they try to solve the engineering challenges of floating living, and in general explain to potential donors or potential citizens why we can and must compete with land-bound governments by undertaking these experimental communities on the bounding main, free as the sea breezes that blow, yet secure from the sea waves that can destroy.

Before they can live on the open ocean forever, the Seasteaders figure, they should try to live on *some* body of water for a long weekend. Hence the Ephemerisle festival, a place for Seasteaders and interested parties to rent houseboats and try to build some makeshift amateur flotation and see what water living is all about.

For insurance reasons, the Seasteading Institute stopped being the technical hosts and leaders of the event after year one; it is now officially just a random, unorganized gathering of people on the water.

The through line from floating party to floating city-state isn't entirely clear, but Ephemerisle already works as an example of the freely organized anarchy of the market providing everything we need, just like Ron Paul promises it does and will. People collaborate on renting houseboats and forming them into a stable U-shape, lashed together. Decentralized taxiing to the site by rented motorboats manages "public transportation," fun and efficient. Food is bought and brought and cooked and shared, not because someone planned it or forced it to happen, but just because people can mostly be counted on to just take care of themselves, and to be generous when the implicit social contract calls for it, and they're with people they have chosen to be around who don't expect to be or insist on being taken care of.

Here on the houseboats for three days, one hundred or so folk, most of them young sybarites from tech and science industries and mostly libertarians, gathered; many of them worked together impromptu to build amateur but very sturdy floating platforms out of readily available or unlikely items, stray plywood sheets and two-by-fours framed around buckets and old plastic maraschino cherry barrels. On those boats and platforms they'd celebrate, mingle, talk their spacey brand of politics, and enjoy the basic prerogatives of free (or yearning to be free) Americans partying in the summer on the water: barbecuing, dancing, swimming, and sliding down a makeshift tube made of giant industrial water pipes.

They also—remember, they're libertarians and techies—spent much of the sunny afternoons both giving and listening to a series of short lectures, splayed out wet or dry on the platform, or sitting on benches quickly cobbled together, the legs made from five-gallon plastic buckets. The range of the talks was sarcastically summed up by one attendee as being "eighty percent about Bitcoin [an artificial computer-based experimental currency] and eighty percent about polyamory."

That was not strictly true, of course, never mind the mathematical impossibility. But it summed up what they care about. These would-be Seasteaders are serious about alternative ways of imagining and structuring society and culture, and that covers everything from money to love. In addition to the social speculative fiction, we got lectures from actual working experts in everything from cellular degeneration to robotics to working in Antarctica to exploring hidden undersea caverns via remote control submarines, sometimes still wet from a swim.

Naturally, such a scene attracted people with serious thoughts and stories about Ron Paul.

Not that you could count on them to care about electoral politics at all. During one of my interviews with Friedman about Seasteading, he invidiously compared the returns libertarians could expect from expending their wealth and time and energy on something as obviously silly and hopeless as trying to get Ron Paul elected president to the more sober, sensible, realistic path he was hewing: trying to build artificial new nations on the sea. Still, a surprising number of the hundred or so people here were Paul enthusiasts, or had a lot to say about him.

One night I and a small group of other attendees huddled in plastic deck chairs in a circle around three leading Ron Paul activists from Los Angeles. People were squeezing and gliding past us through the entire interview, moving from the back deck of the houseboat to hop out on the jerry-rigged platform that hadn't existed twenty-four hours earlier, to dance to the music projecting from the back of another houseboat, with a full professional DJ setup. All unorganized, all voluntary: one Seasteading employee told me that events like this are more than just a party, more than just an indulgence: "It's a stirring example of why I have faith in humanity, in the idea that people can and will solve their own problems" if left free to do so.

I was talking to Roger Pruyne, Jon Arden, and Stanton Cruse, who all spent most of their free time and energy in 2007 and 2008 trying to spread the word that Ron Paul ought to be the next president of the United States. They made dedicating their months to a quixotic campaign to get an old obstetrician the Republican presidential nomination sound even more fun than this water party.

Pruyne, a senior Web developer for Rockstar Energy Drink, and Cruse, an animator and restaurateur, already had that key element of a Ron Paul fan's DNA: a keen interest in outré and forbidden ideas. They were part of an Internet community dedicated to an obscure energy source called "hydrogen on demand," involving splitting water and using the hydrogen produced to help power cars. Cruse's politics were standard neocon then—"I thought it was great we were democratizing the *shit* out of Afghanistan"—but encountering Paul videos on YouTube turned him around. A group of young Hollywood professionals, intellectuals, and hipsters built up around them, and they all grew to share a passion for Paul.

Cruse joined up with Ron Paul fans' efforts to get on the Los Angeles Republican Party Central Committee. He went down to a local convalescent home to get some Republican voter signatures for his effort, and one old man said to him, hey, you're a Ron Paul guy—how do you feel about gay marriage? Assuming he was dealing with a cranky old right-winger, Cruse "started giving him the standard, you know, 'marriage is traditionally between a man and a woman . . .' and he said, 'You don't have to say any more, I'm not all that conservative when it comes to those things.' I thought, man, I just lost a signature; I should have just given the actual Ron Paul answer, which is, it's none of my business or anyone else's. I went back and saw him later and he signed for me, saying 'I hope you learned a lesson, young man.'"

The three of them and their Ron Paul buddies in Los Angeles helped run a grassroots Ron Paul call center. They'd robocall messages about Paul that gave a number to call for more information. The guy who ran it, Ben Ahdoot, worked in real estate sales. In a Ron Paul irony, the housing downturn caused by the Federal Reserve–fueled speculative bubble meant his big office space on Wilshire Boulevard was nearly empty—indeed, the whole building was almost empty—and he gave the space over to the cause of Ron Paul phone banking. Sure, getting automated political calls often pisses people off; some people would call back just to vent that anger; but many called back to say they were annoyed by the call until they realized it was about Ron Paul.

Soon two entire floors of the largely empty parking garage in Ahdoot's office building had Ernie Hancock-trained activists stenciling billboard-sized Ron Paul Revolution banners. They'd make signs all night, hang them just before dawn, then go to work. "I used all my sick days, all my vacation days, doing this kind of shit, went through all my savings," Arden remembers. They'd see other people doing the same thing, and not even recognize them. This was getting huge. It was bigger than them. Arden swears that at least twice cops totally saw him hanging Paul banners in illegal places and chose to let it go. He saw a commenter in a YouTube Ron Paul video thread asking, "Did anyone see that sign on the 405 saying 'YouTube Ron Paul'?'" Arden beamed with pride. That one was his.

The boys ran regular weekly meetings, which they called Ron Paul Hollywood Meetup, which morphed to a general pro-liberty meetup after the campaign. There was a self-help and educational end to it all—eventually they had hooked up with a larger group of young libertarian entrepreneurs in the Los Angeles area. It was more than a political hobby—the liberty movement, the people they met and friends they made and obsessions they pursued, became their life.

Pruyne rented a hip loft space at Hollywood and Vine, ripped out the carpet and built a stage, and made it a general Ron Paul/libertarian groovy hangout space. Their meetings became invitation-only as they got tired of some of the weirdos— including some who they began worrying might be provocateurs, a guy who claimed to be a former cop talking bullshit like, Hey, making videos isn't good enough, I wanna *die* for my country. A Japanese medical student showed up, having flown himself to America just to attend a Ron Paul Meetup. And their buddy Bill Johnson—who, they all insist, exhibited no signs of any weirdness, and just had a nice house in which he hosted high-ticket Ron Paul fundraisers and some of their meetups—ran for a Superior Court judge position in Los Angeles, as a Ron Paul guy, and then was exposed as having been in a past life "Daniel Pace," who wrote books advocating that only descendants of Europeans should be allowed to stay in America. However strange the world around Ron Paul got, Ron Paul himself never let them down.

The next big thing after Ron's campaign ended was "End the Fed!" The Los Angeles gang organized five-hundred-person rallies around that cause, spearheaded by local Ron Paul activist, financial analyst, and yoga teacher Steven Vincent. (The rallies would typically end with a small sidewalk bonfire of the Federal Reserve's vanities, its paper money.) By 2011, people in their crew were running huge liberty-oriented events like "Nullify Now," sponsored by the Tenth Amendment Center, dedicated to states' power to override or ignore tyrannical federal laws, not explicitly Ron Paul–themed, and selling hundreds of *tickets*. "Ron Paul ignited a vibrant pro-freedom movement that exists independently of him," Pruyne says.

While there ain't no such thing as a free lunch, there is such a thing as a cost-shifted one, so later on during Ephemerisle I wandered over to find representatives from Seasteading funder Peter Thiel's operations sharing deliciously marinated meats. (Thiel

himself, though his presence was rumored, didn't make it.) Jim
O'Neill was quietly and undemonstratively warm and welcom-
ing to all the strangers hopping onto and wandering through his
boat. O'Neill works with Thiel on both his philanthropy and his
hedge fund Clarium.

While working outside politics now, O'Neill has his own
political past. He was a speechwriter in the second Bush
White House (not for the president himself), and now follows
Ron Paul with interest. O'Neill's political experience fed his
libertarianism—working in government helped him realize that
government as practiced is neither virtuous nor realistic. He notes
that "it's striking how world events have conspired to dramati-
cally confirm Ron Paul's views on money and war. What events
could have turned this implausible candidate to a real one? Well,
maybe a bank collapse followed by a big government bailout,
with the U.S. invading three nations, Gitmo still open, we keep
torturing, threatening whistle-blowers with jail." After all that,
Ron Paul's vision of the nature of the U.S. government and its
problems now seems the only realistic one.

After encouraging me to try the makeshift slide—it was great,
but the water at the bottom was too sharp and chill to enjoy for
more than a frantic few seconds of swimming—Zoe Miller, who
came in with her kids from Reno to enjoy Ephemerisle, shared
some stories about the frustrations Ron Paul activists had with
the GOP powers-that-be in Nevada in 2008. When the state Republican Party calls her asking for donations, she told me, she
delights in upbraiding the poor mendicant on the other end of the
phone for the party's maltreatment of her man Ron.

I was reminded that the consumer surplus of the Paul movement mostly goes into the pockets of his fans, not the professionals trying to run a winning campaign. The richest joys of the
Paul Revolution come from the opportunities created for people
of like minds to gather about matters that you can belittle by call-

ing "political" but that nonetheless touch on the deepest parts of people's sense of sociality, their sense of who they are. It creates a space for people who see life the same way to be together and play, in a context that feels simultaneously important and fun. The Paul Revolution, metaphorically and sometimes literally, allows for dancing, and that's a big part of what sustains it.

L ibertarians are peculiar for many reasons, one of which is how rationalist their political thought tends to be. They have a set of first principles about ethics and politics from which their political positions derive with what they see as inexorable logic. That means many people who love libertarian ideas abstractly expressed—life, liberty, and the pursuit of happiness are great!— flee when the remorseless logic of believing that government should only restrict our actions when they directly harm the life or property of another leads to the conclusion, say, that forcing kids to attend publicly funded schools is illegitimate.

The positions of the two major parties are indeed just a clumsy amalgam of attitudes, protecting the interests of perceived constituents and arising from historical or sociological accidents. What is the logic that unites, say, defense of teachers' unions, belief in universal government-supplied health care, and gay rights? Or prayer in school, low taxes, border walls, and being *against* gay rights? Libertarians' positions, on the other hand, actually make sense down the line, derived from specific principles about government's purpose and legitimate powers.

Ron Paul since 2007 has seen his fan base expand beyond that circumscribed world of tight, sensible, readily deducible conclusions. Many of his new fans—overwhelmingly *not* derived from any preexisting self-aware mass of libertarians—just like something about him.

Conservative journalist Tucker Carlson was enough of a fan

of Ron Paul to have emceed at Rally for the Republic (though not enough of one to actually stay through the whole event once he started hearing 9/11 Truth talk from Jesse Ventura). In September 2011 on the Fox News comedy talk show *Red Eye*, Carlson was giving some public love to Paul. At the same time, he was spelling out explicitly that however much he respected Paul, he was *not* going to vote for him. He noted that while millions say they love Paul, when you break down the full spectrum of his firmly held radical positions, he didn't think there could be more than nine people in America who *really* agree with the totality of Ron Paul.

This is not literally true, but it hits on an interesting aspect of the Paul revolution for Paul's old libertarian fans. The politically savvy understand that politicians can win over voters who don't agree with them on every issue. It feels different with Paul, though. He exudes so much coherence and integrity that people sense it *matters* when they disagree with him more than it matters when they disagree with a politician who is just a grab bag of acceptable stances. When someone disagrees with Paul, they realize it means he really isn't the guy for them, even if they agree with him on a lot of issues.

Many fans love the integrity they sense, the respect for their liberty, so much that they often decide that they agree with Paul even if they don't necessarily understand his reasoning. I've had a handful of Paul fans tell me that they assume Paul is correct on something even if they haven't thought it through themselves yet. This is not out of cultlike devotion. It's because they see that he's an intelligent man whose thought process they trust on the areas where they do fully understand his reasoning.

Sometimes Paul fans decide that Paul must agree with *them* even if he really doesn't. When a Paul enemy challenges them with some wild libertarian position of his that they think no sane human can defend, they get angry. Why do people say Paul believes in *legalizing drug use* or *thinks it's okay if Iran has nuclear bombs!* I've heard more than one very devoted Paul fan tell me a story like that, outraged on Paul's behalf. (The "okay if Iran has nuclear bombs" point is somewhat ambiguous; Paul has said directly he doesn't want anyone to have them. But when people accuse him of being okay with Iran having nuclear bombs, of course they mean "the U.S. shouldn't start a war to prevent Iran from having nukes," and Paul does believe that—he believes that, as with the Soviets, we can be capable of containing the dangers of a nuclear power without resorting to open war.)

Paul does believe those things, as crazy as they sound to most Americans and as unhappy as an average Paul fan might feel having to explain or defend such stances to, say, their mother. This is why Paul hews to the Leonard Read vision of ideological change: there are still minds to educate on the full implications of personal and constitutional liberty, and he and his organizations are working on that. Paul got caught up at a debate in 2011, being asked how he'd get his budget concerns through a divided Congress. He needed to say that a world where enough people liked Ron Paul enough to propel him to the presidency would be a political world in which Congress—usually a lagging indicator of the people's will, not a leading one—would take Ron Paul's budget seriously.

Before taking Ron Paul's budget seriously, Ron Paul's fans and Ron Paul himself need to be taken seriously, and they aren't always. Pure ideological prejudice and confusion are at work, but there are other reasons, including that Paul tends to attract a type of fan whom everyone is already primed not to take seriously, regardless of their beliefs on, say, spending and tax priorities—the conspiracy theory folk.

It's no mystery why people worried about, say, the Bilderbergers and the Trilateral Commission, if they have a favorite political candidate, would glom on to Ron Paul. Many Americans have conspiratorial nationalist tendencies, and Ron Paul, unlike most politicians, understands them and is not afraid of them. In attempting to win their support and their donations over the years he has chosen to speak their language at times. On a more basic level, he's the only national political figure who is willing to grant a key element of the conspiratorial worldview that marks them as verboten, and often marks Paul as verboten: that the United States government doesn't always mean well, isn't always a force for good; that in fact, domestically and internationally it does evil things that deserve to be condemned. Liberals believe in the government's inherent good too much; conservatives theoretically see problems with the state but are too jingoistic about their own government to highlight its international or national wrongs. That leaves Ron Paul to say the emperor not only has no clothes, but is ugly to boot.

Phil Blumel, the man who organized the event in January 1988 where I first met Ron Paul, has been a Paul fan and supporter ever since. Blumel has been active in Republican Party politics in South Florida for more than a decade now. He came to Paul from a Libertarian Party background. Despite having gone as a teenager to a Bircher youth camp along with his miscreant best friend and punk rock bandmate, he doesn't dig the populist conspiratorial nationalism stuff and doesn't think it serves Paul well at this stage. Paul seems to have come to the same realization and doesn't talk about the Trilateral Commission or the North American Union as part of his presidential campaign.

That stuff comes from an outsider's attitude, and it has no place culturally in mainstream American politics. As Paul so often says from the campaign trail in 2011, the mainstream has started to come to him. But he still scoffs good-naturedly at the idea that *Ron Paul has become mainstream*. Some form of oppositionalism is still, as he knows, key to his appeal. But there is oppositionalism, and there is oppositionalism. Being the antiwar, anti-drug-war, don't-just-audit-but-end-the-Fed GOP candidate is quite enough oppositionalism for one candidate to carry.

But to the frustration of people who either want to humiliate him and force him to heel, or who admire him but are put off by conspiratorial thinking, Paul will also not condemn or turn on those ideas or those people. When accused of Bircherism, he'll just explain why he sees no disgrace in either being a member of the John Birch Society or being loved by one. He even told the *New York Times*, not the most understanding audience for pro-Bircher sentiment, "Oh, my goodness, the John Birch Society! Is that bad? I have a lot of friends in the John Birch Society. They're generally well educated, and they understand the Constitution. I don't know how many positions they would have that I don't agree with. Because they're real strict constitutionalists, they don't like the war, they're hard-money people."

It seems a curious disconnect to believe in the power of the Trilateral Commission, Council on Foreign Relations, and Bilderbergers, or conspiracies even darker and unnamed, to plan and shape American politics and to believe it matters how hard they campaign for Ron Paul. A handful of Paul supporters either told me they believed, or asked me if I believed, that Paul faced obvious physical dangers if it seemed he might actually win the presidential nomination. Believing this didn't seem to dissuade them from working to create the very conditions they feared might lead to their hero's death.

When I asked Paul why he appears on the Alex Jones radio show (Jones is the leading all-purpose conspiracy theory promoter in America today), Paul plays with me a bit—why am I suggesting there might be something wrong or mistaken about him appearing on Alex Jones's show? "Because he says some things that are wrong, that I don't agree with? Well, that might exclude me from every national TV program. I mean, I get on those shows and they are pushing me on why they love assassinating American people, bombing foreign countries and war." The other opinions of the media he appears on, he says, do not "bother me one bit. I speak for myself, not for the people talking to me, whether it's Jones or any national TV media whose views are much more dangerous than Alex Jones. It's not a mistake to talk to people, even if you don't know or don't agree with all the details of their thinking. If you asked me to write down the good things or bad things about Alex Jones I wouldn't be capable of doing it. We talk about the issues I want to talk about." And getting to talk about the issues he wants to talk about, to as many different people in as many different places as he can, is why Ron Paul does what he does.

Paul's ecumenism on these matters enrages his enemies and frustrates some of his friends. Slattery from the New Hampshire campaign in 2008 laments that "Ron won't tell them they are crazy and tell them to shut up when they confront him with these 9/11 Truth questions. He's never mean or dismissive." Daniel McCarthy, who worked briefly in Paul's communications operation for the 2008 campaign, says that "my impression was that though some percentage of Ron Paul supporters were interested in the 9/11 Truth movement, even those wanted to 'Go Clean for Gene' and not make spectacles out of themselves, since they knew it was used against Ron."

This vague sense of "association" with people with outré beliefs that Paul himself has never directly embraced is not Paul's only unique problem as a national political leader. Another is his strong tendency to lead his supporters away from electoral politics altogether.

If you were attracted to Paul and his message, and were intelligent and curious, and your soul resonated to the notion that there is something untoward, even sinister or evil, about using violent force to shape social relations (and you began to understand that violent force, really, is what government is), you probably started

reading some of the things Paul pointed you to, and then kept pil-
grimming down a long, meandering, complicated road of ideas.

You were likely to find your thoughts progressing from believing that the federal government should restrict itself to its constitutional functions to realizing, via nineteenth-century abolitionist
and anarchist Lysander Spooner (a libertarian favorite whom Paul
gave a shoutout to in his 2011 book, *Liberty Defined*) that, hey,
I never signed this Constitution and I don't know anyone who
did. So how can we say it's a binding social contract that requires
us to obey the rules a small gang of politicians and bureaucrats
dictate—politicians chosen through a process that seems just like
a secret band of brigands electing a leader, that "advanced auction
on stolen goods" that H.L. Mencken joked about? Who could feel
good about being involved in that process?

But all wasn't lost. You might have come across the ideas of
libertarian movement theorist Sam Konkin along the way, or
someone who was influenced by Konkin. Or maybe your pals at
the anarchist bookstore or those crashing on your couch in their
perambulations around the nation helped you see that there was
more to life than the political process: there was the *agora*, the
market.

That didn't have to mean a place where people bought and
sold things to each other with cold mercenary calculation. It
was the wide world of free choice and free trade on any terms,
with anything, using any currency, as long as everyone involved
wanted to be involved. That was a rich world indeed; it brought
you pirate radio and underground clubs and circles of mutual aid
and cheap or free places to crash, and cool under-the-table jobs,
and it was a hell of a lot easier, even in an increasingly TSA'd and
Patriot Acted and Real ID'd America, to find freedom by just
grabbing it wherever you could than it was to find freedom by
trying to get someone elected president.

Many of the eager young Ron Paul revolutionary armies of

2007–2008, while they still bore residual affection for the man and movement and might sport their old rEVOLution T-shirts or flair, couldn't be bothered to do anything active to promote him this time around. At a Ron Paul picnic in Los Angeles in July I met Jason Wohlfahrt, who told me that he and many of his younger friends energized by Paul in the last campaign had decided the electoral process is hopelessly corrupt or irrelevant to either their lives or to forging needed social change. "Post-political" is how he describes himself. Agorism has won his heart. Agorists don't see any point in making a frontal assault on the state, much less being naïve enough to believe that elections created by the state will lead to any change a libertarian or anarchist should care about. "If voting could change anything, it would be against the law," as the old slogan goes. You get rid of the state by ignoring it, not fighting it—or at least get rid of it to the extent it affects you.

Wohlfahrt speaks for a common disillusionment on the part of Paul activists, particularly the more intellectual-radical end, with the aftermath of the 2008 campaign. He was especially annoyed how the energy and direction were aimed toward Campaign for Liberty, involving "funneling money into things we don't believe in." (Three hundred and fifty thousand dollars of dedicated CFL money, given to CFL explicitly for that purpose and so, CFL fans argue, not coming per se out of general activist support, went to create an ad that activists thought was tantamount to an endorsement for a 2010 Senate candidate from Colorado who suggested we should not be withdrawing from Afghanistan immediately. The Paulista grass roots tore their hair out in rage.) Wohlfahrt was quite serious when he told me that various friends from Texas totally ruined their professional and personal lives giving everything up for the promotion of Ron Paul. A few years later, being a missionary for Ron Paul started to seem to many young libertarians like not the most libertarian thing a young, impassioned activist could do with his time.

Ron Paul, in the old intramovement debates during his 1988 Libertarian Party run, was supposed to represent the bourgeois side, and in his personal life he did. But when he became the national lightning rod for antistate thought in 2007, he began leading lots of youngsters to the world of ideas and attitudes displayed on the LewRockwell.com website, from the explicitly antipolitical to the support for alternative-lifestyle diet concerns and an overpowering sense that a statist culture was well worth trying to evade or escape. Paul has thus become the apostle of libertarianism to a new breed of people who live "left" lifestyles, but with a free-market flavor.

Tucker Carlson noted this peculiar aspect of Paul's fan base to me. "People see in him what they want to see, but they see more than anything a rejection of conventional politics and conventional ways of thinking. Declaring you are for Ron Paul, going to a Ron Paul rally, it's another way of saying that you are willing to completely rethink the way you see the world. It's like a declaration of personal identity," Carlson says. "You know, people don't vote for a candidate because they agree with the candidate. You vote because voting for that candidate makes them feel good about themselves. That's a basic truth about politics lots of people don't understand. But with Paul, it's this especially weird mismatch of counterculture people, drug people, people who just don't buy into any of the core assumptions of American life. And their leader is like Ward Cleaver."

But, Carlson notes, Paul's radical fans are well earned by his radicalism. "When I hear people say that people want to be free, I think, do you have any evidence for that at all and do you know anything? And the answer is no on both counts. If you had a society structured on the lines Ron Paul espouses—and I'm for it! I get what he says and I'm for it!—if we had a society like that, it would be so much more different from the society we live in than Paul fans understand; it would be a completely different society. Ron Paul is *genuinely radical*, and I say that with some admiration."

Vijay Boyapati, who upended his life and sacrificed perhaps as much as anyone for the Paul cause when he launched Operation: Live Free or Die in New Hampshire in 2008, learned some genuinely radical lessons that don't necessarily bode well for the energy of the Paul campaign moving forward. "The experience made me cynical and radicalized me," Boyapati said four years later. "I became a more hard-core libertarian. It was good to see the mechanics of grassroots democracy up close and personal, but to me it was very disappointing. I helped create a grassroots movement, and I had the idea I could bring a thousand people out to New Hampshire—it ended up around five hundred to six hundred—and that if I could knock on every door and tell everyone about Ron Paul, they would magically go, 'Oh my God, this is obvious, why didn't I know about this, why didn't I know about him?'

"It turned out that, first, you can't convince a person in a three- or five-minute meeting to change their worldview. It's just not gonna happen. I also realized that grassroots politics plays a secondary role to much bigger forces which swing the course of elections. Whether Ron Paul was in the debate and what treatment he got made a bigger difference than me knocking on doors. If a moderator in a Fox debate characterized him as a fringe person that you should never vote for, that shaped more than grassroots campaigning."

Still, Boyapati had no regrets: "The experience itself was one of the most exciting of my life. Meeting all these libertarians with such passion—people came from all over the country; one woman packed her family, she was pregnant, and brought her two children and drove all the way from Arizona to New Hampshire, just put everything in the car, didn't have any money, but just wanted to tell people about Ron Paul and his message. That made it worthwhile: getting in contact with so many libertarians and feeling their passion and hearing their stories—even if our effect was marginal at best."

By asking Paul face-to-face where he got his ideas about economics from, Boyapati was led to Mises, and the Mises Institute, and Rothbard and Spooner and the people who taught him that electoral politics isn't how to build a free society. He's adopted the "voluntaryist" perspective that questions the morality of electoral politics in general. (Even if you can argue that voting for Ron Paul is legitimately defensive—he's going to lessen the power of the state, not increase it.) Carla Gericke of the Free State Project in New Hampshire tells hard-core voluntaryists—she's surrounded by lots of them—that voting for Paul in the *primary* is ethically defensible even under the strictest anarchist premises: you are not voting for him to take an office, like the U.S. presidency, that inherently exercises immoral coercive power. You are merely trying to select him to a specific position within the Republican Party, one that is not in and of itself a political office.

Boyapati used to think of himself as a constitutionalist, but Spooner disabused him of that naïveté. "If I could speak to the old me, I'd tell him, some stuff you believe isn't really true, and here's why. I'm convinced by Spooner's argument that the Constitution is an illegitimate contract. I've become much more sensitive to certain types of injustice. It makes it harder to watch or read the news. It's upsetting to read every day the generally accepted view that something is correct or should be done, like that war is legitimate, or the bombing of another country is acceptable." It all makes this former news junky avoid the news of the world around him, Boyapati says. It's all just too . . . *wrong*.

While the antipolitical likes of Wohlfahrt and Boyapati are not unique among the troops in the aftermath of the Ron Paul Revolution, they aren't everything. Standard political and social rallying to shout the name Ron Paul to the heavens is still happening; I went to one such rally early in the election season, in early

July 2011, in a park in Studio City in the San Fernando Valley, north of Hollywood. (It's actually where I met Wohlfahrt—he still likes such events for their social aspect.)

The event was wrangled by Steven Vincent, an intense leader of the Ron Paul movement in the Los Angeles area, who was a key thinker and actor in launching the notion of Tea Party rallies (he organized the 2007 version on the Santa Monica Pier) and pushing a national "End the Fed" street protest movement. He's both a technical market analyst and a yoga instructor—a living example of the Ron Paul movement's melding of the economistic and the groovy. Vincent quickly told me of the government failure interfering with our day. He had done all the right things to reserve the park from its government owners and showed up to find a family had been rented the same space for a kids' birthday party. A bounce castle loomed close to the tables for Oathkeepers (a group of law enforcement agents and military who vow to never obey orders requiring them to do unconstitutional things, founded by a former Paul staffer, Stewart Rhodes) and John Birch Society booths.

Vincent is an intense guy, lean, with short-cropped dark hair and Roman nose, and he is here to bring to the Paul people the good word, and the good word now is "global debt repudiation." He has been the apocalyptic guy, and he doesn't want to be anymore. He doesn't think the Ron Paul movement collectively should be that guy, either. Sure, the economy is facing some problems, but with the loosening of the strangling hand of regime uncertainty that's keeping capital held tight for fear of what crazy confiscatory thing the government might do next, he thinks we could be on the cusp of an era of economic growth to match the nineteenth century, with new energy and tech initiatives. But there's something we have to do first: global debt repudiation.

He's enthusiastic about it one on one, and he was enthusiastic about it while addressing the seventy-five or so Paulites gathered on a typically bright and clear Southern California summer Saturday. The Greek debt crisis was in the news that week; the country clearly couldn't make good and it looked like the government was going to just squeeze its citizens hard to try to stay afloat. Vincent saw this as the fate of everyone in the West: driven to penury by their governments because of the governments' sovereign debts, and the governments' refusal to allow banks to take the hit for debts gone bad. Vincent is a great spellbinder, and he had a compelling rap/chant for us: "We didn't borrow it. We didn't print it. We didn't spend it. It's not our debt and we won't pay it. We will not give the sweat of our brow to pay it off. We repudiate the debt!"

Default, he insists, reflects the real market situation. We need to force a reset in the global economy. The American people did not choose the behaviors that led to U.S. government debt, and they should not be mulcted to make good on the politicians' idiocy: this needs to be a key part of the Paul movement message. (Paul did talk about private debt liquidation in his campaign talks that season, but not government debt repudiation.)

Vincent was trying to be a good motivational speaker. What can we do, what can I do? Paul fans wondered. Well, "find the thing you enjoy, and then dedicate yourself to it. Whether it's emails, signs, knocking on doors every week, whatever it is, do the hell out of it. There is no way they can stop us. We will win."

Will they really? That presumes it's just a problem of message saturation—that if enough people hear about Ron Paul from enough different places, they will realize they agree with him and vote for him. But Tucker Carlson may have been onto something: the mass constituency for no war, no Fed, no drug war, no regulatory state, and a planned drawback of the entitlement state may not be as big as Vincent thinks.

"You know, there's something I really miss from the last go-round. I say: 'Ron Paul!' You say: 'Revolution.'" We did. "Do a few 'End the Fed's!' It feels good! Opens you up in here!" He pounded his chest. He is a yoga coach. "It's been stuck up in here for months. Let it out! We're all about to be breathing again."

The mood at the picnic was not manically optimistic, but it was determinedly so. There is no demographic that should be impervious to the power of Paul, I'm told—inner cities need to hear a message of an end to drug wars and overregulation.

I met at this picnic one example of what the Ron Paul Revolution needs to thrive—an accomplished adult, with the "right" background, willing to run for office. His name is Rick Williams, a high-end Los Angeles lawyer with a classic aging Western hero look. He could be cousin to Ronald Reagan—a politician he admired in his time, but whose foibles and weaknesses he's since come to see. Williams is also willing to loudly violate Reagan's famous Eleventh Commandment of politics: Thou shalt not speak ill of a fellow Republican.

Williams and I met at Mel's Diner on the Sunset Strip in Hollywood. Williams first encountered Ron Paul during the 1990s, at a meeting of the paleoconservative-paleolibertarian fusion club the John Randolph Society. Williams is a lifelong Republican who found that Goldwater's "ringing affirmation of freedom really touched me" in his youth. He worked in the law firm that Nixon retreated to after his 1960 defeat. He drifted to Pat Buchanan in the 1992 and 1996 elections; Buchanan was the only one who got what a bad idea the first Bush's first Persian Gulf War was.

Williams is a religious man, but he got so incensed by how warmongering his church brothers and sisters became after 9/11 that, even though he was a deacon, he stopped going to church for a while. He sees Paul's place in history in biblical terms: "A lot of people compare him to Jefferson, but I think the better comparison in my mind to Ron Paul is to Moses, who leads his people out of slavery to the brink of the promised land but never enters the promised land himself. He's a living Moses who has taken his people to the brink of what I think is a major epic change."

Williams puts his time and energy into alliances with different groups of young, media-savvy Paul fans. They use online media, entertainment, and information of a roughly Paulite bent, first under the name "Break the Matrix" during the last campaign cycle, now "Revolutimes." Williams took a year off from his legal work a couple of years back to work on a book about the damage that bankers in collaboration with government have done to the economy. He sees us skidding wildly into an economic crash of frightening proportions—but he doesn't think that's necessarily such a bad thing. But that doesn't mean he thinks there's nothing that can or should be done. He ran for a seat on Los Angeles's Republican Party Central Committee, heeding Ron Paul's advice that the best thing his supporters could do was reshape the GOP in their image, from the bottom up.

Some of Williams's friends in the Ron Paul world had already won county party positions in 2008; he waited until 2010 to run, forming a "Freedom Slate" with six other candidates—not all Ron Paul people, but all freedom-oriented candidates who see the need to reshape the GOP so it is more than just the second party of big government.

In the Republican Party but not necessarily of it, Williams quickly showed the side that made the existing party faithful leery of him: "The Republican Party as it exists is a massive failure and will continue to be until there's a change of direction. I ran as an anti-Republican establishment candidate. I think the Republican Party is a disgrace. As a lifelong Republican it makes me sick what the party stands for, who they nominate, who runs the party."

It's not hard to see why this message might have been a hard sell to GOP primary voters in the Los Angeles area in 2010, whose heads were full of the desire to send Obama a message—though Williams insists he'd be equally bold when talking to local Republican women's clubs. "I did not get pushback from local Republican voters," he says. "I got loud applause. I did get pushback from establishment Republican Party people who are the current officers and supporters of the Republican insider group in California. They were very much against me."

His Paul brethren ran into that kind of trouble even after winning their seats. Robert Vaughn, a Ron Paul and Campaign for Liberty activist and a local child protective services worker, won a seat on the Los Angeles GOP central committee and then was elected as a vice chair of the county party's board. He ended up chair when the people above him stepped down. The party hierarchy didn't like this and robbed him of his title in a parliamentary coup.

Although Vaughn spent six figures fighting, and used Williams as his lawyer, he ended up losing, the court basically declaring it an internal matter for the Republican Party. The existing powers of the party, Williams says, have a very particular vision of the Republican brand that they will lie, cheat, and steal to protect. "Limited-government Republicans, and antiwar ones who believe we should downsize our overseas military presence, are not the brand the Republican Party wants to sell" in Southern California. (Or anywhere.)

For most of us, what defines the Republican Party in operational terms are the candidates running for and sitting in office. It isn't instantly apparent why Paul and his people stress the slow, dull, unsexy taking over of party offices on the local level as key to the revolution. But that strategy is exactly right, Williams says.

"The most important thing the Central Committee does is register and recruit voters and encourage, locate, and select candidates. Those two functions define who the party is. The candidates you see running for office [are usually] candidates that the Republican Party encouraged to run, promoted, and helped. So even though the leadership is faceless and largely ineffective over the last twenty years, it has a very important purpose. Also, they are the people who become the delegates to the Republican

National Convention, and they become the people who vote, or decide who the Republican presidential candidate is."

So Williams is planning a 2012 run for one of California's U.S. Senate seats, up against Dianne Feinstein. California last year changed its primary process: now everyone gets to vote for anyone in the primary, and the candidates can identify themselves by party affiliation or not, as they choose. The top two, regardless of party, contend in the general election. Either two Democrats or two Republicans can end up being the only choice. How does Williams intend to sell himself, in the first experiment with this fresh process? "As a Ron Paul revolutionary, with the emphasis on *revolution*."

Won't that scare off Republican voters?

"They need to be scared. If they are not scared, they should be." The sanguine apocalyptician in Williams came to the forefront. "Just look at what's coming, the disaster that's been created by the combination of big government and big banks. If they are not scared of the disaster coming our way, they are not in touch with the world we live in.

"We're not talking about something solvable through the political process, like what Obama and Geithner were trying. The crash is gonna come. The debts have to be wiped out and repudiated no matter who the president is. The issue is not 'Can we save ourselves from the coming collapse'—which will be huge!—but what will emerge from it, what the new country, the new world will look like. I don't offer a solution to the coming collapse. There aren't any. I see the need to shape the new world we are going to live in.

"If you had a big dinosaur coming at you, and you shot it in the heart with an exploding bullet, it would take fifteen minutes for the dinosaur's fingers and pea brain and toes to realize that it's dead," Williams concluded. "We are living in those fifteen minutes."

★

CHAPTER NINE

RON RUNS AGAIN

The demand for Ron Paul to run for president again was clearly there as the 2012 campaign loomed. But would he obey the free market in ideas and supply it? The first big public sign of still-lively Paulmania was the Conservative Political Action Conference in February 2010, when Paul handily won the straw poll of the most dedicated young conservative activists. Like Obi-Wan Kenobi, the GOP's slaughtering of the dream of President Ron Paul in 2008 did not kill it. It made that dream more powerful than the GOP could possibly imagine. Paul racked up another CPAC poll win in 2011. The Cato Institute's David Boaz recalls noticing that the bulk of the crowd rose to their feet in cheers at nearly everything Paul said; the only section of the crowd he noticed keeping their seats, it turns out, was where the organization's older, big-donor bigwigs sat. This was still very much a revolution of the young.

In May 2011, Paul made official what everyone guessed: he was running for president again, having decided that citizen awareness of the bloody mess of the out-of-control welfare/warfare state might be hitting critical mass. Then on July 12, Ron Paul surprised his fans by announcing that this term in Congress would be his last. He had also recently sold his house in Lake Jackson, Texas. "He's leaving his job, he got rid of his house—is he expecting to be lifted bodily to heaven at the end of the campaign?" joked one old Paul campaign worker. A congressional staffer, caught by surprise like the rest of the world by the announcement, was humorously despondent: "I either sell out and go work for Michele Bachmann, or move to New Hampshire and smoke dope and get arrested."

Paul himself told CNN that night, "I don't think I have a political career so much as trying to change the course of history." Norm Singleton says he and Paul were joking after the announcement, "about him telling [House Speaker John] Boehner that 'now that it's my last term, I can finally vote my conscience!' "

While leaving Congress was a surprise, it probably shouldn't have been much of a shock. Ron Paul never really enjoyed the actual experience of being a congressman. "I expected it to be bad. I didn't expect to accomplish much," he says. "People ask me how I can stand being in Congress and not be frustrated. I just had low expectations. I figured things would just continue the way they are, and that way any victories would be pleasant.

"I thought my role was to speak out about issues I thought were important and then found, wow, you can get elected doing this! Then my next role was to vote exactly as I said I would, follow the Constitution as precisely as I can, and see if a person like that can get reelected. I did that and figured I'd just plod along and try my best to set a record that someone, someday might read. Like I'd look at H. R. Gross's record."

As his old aide Bruce Bartlett said: "Being in office is strictly a means to an end for Ron. I asked him directly last time I talked to him, 'Why do you keep doing this?' It was several years before the '08 presidential run. 'You don't have much in the way of legislative accomplishment.' And he says, he likes it, he gets to say what he thinks and meet interesting people, and that was enough. And I believed him."

With Paul a newly minted national hero from his presiden- tial run, and a world of new national institutions such as CFL and YAL sprouting up around him, and his status as a national spokesman for liberty ideas secure, he no longer needed the con- gressional berth to get what he wanted out of national politics to begin with. He and his LibertyPAC can feel freer to fund liberty- oriented primary challengers to other Republicans. And besides, he had a presidential race to run. Singleton analogized Paul to Blaxploitation movie hero Shaft—"Now that he's off the police force, it makes him more dangerous, not less." As anarcho-Paulite Anthony Gregory of the Independent Institute said, "The fact that he's a congressman was the only mark against him."

Paul certainly started this run swinging. At the very first de- bate in early May in South Carolina, the media began playing their usual game of "make the libertarian look dumb"—alas, having actual political principles means they can call on you to explain them; no other candidate ever has their core beliefs challenged like this. But Paul was asked about legalizing heroin. And he gave a bravura answer that never once explicitly said, "I think heroin should be legal"—while demolishing any reason you might have for thinking it should be. It's a great Paul ramble; the sentences don't always follow with perfect grammar and logic and structure, but the meaning comes across.

"My defense of liberty is the defense of their right to practice their religion, and say their prayers where they want and practice their life. But if you do not protect liberty across the board, it's the First Amendment type issue, we don't have a First Amendment so that we can talk about the weather, we have the First Amendment so we can say very controversial things. So for people to say that, 'yes, we have our religious beliefs protected' but people who want to follow something else or a controversial religion, you can't do this, if you have the inconsistency then you're really not defending liberty. . . . But yes, you have a right to do things that are very controversial; if not, you're going to end up with a government that is going to tell us what we can eat and drink and whatever.

"It's amazing that we want freedom to pick our future in a spiritual way but not when it comes to our personal habits. . . . If I leave it to the states it's going to be up to the states, up until this past century for over a hundred years they were legal, what you're referring is 'You know what? If we legalize heroin tomorrow, everybody is going to use heroin.' How many people here would use heroin if it was legal? I bet nobody would! . . . 'Oh yeah, I need the government to take care of me, I don't want to use heroin so I need these laws.' . . ."

He got fervent applause. The other libertarian in the race, New Mexico governor Gary Johnson, who suffered the 2008 Paul media and establishment blackout to the nth degree (and then in December gave up on the Republicans, like Ron Paul in 1988, to seek the Libertarian Party's nod), gave a softer answer about merely legalizing marijuana. Ron Paul was running for president again! And his fans rejoiced.

Iowa's Ames Straw Poll is the traditional first chance in the election season for the candidates to prove their organizational mettle. It's as unscientific as can be, and its predictive power for eventual winners is small, but it does have the power to break a campaign, if not make one.

Iowa, for a visitor, is soft, sweet, and bright, light blue and cool white and huge and endless. I was just visiting. I don't really *know* Iowa. But I like it. Not enough shopping for this Los Angeles boy—something as simple and obvious and necessary in my California life as grabbing a soft drink or other caffeine alongside freeway off-ramps is hard as hell in Iowa. And the size means a mere reporter in a car can't manage to keep up with a candidate using a private plane. Paul spent $1 million in the third quarter of 2011 on private charter planes; with his artificial knee, and his love of privacy and liberty, he hates Transportation Security Administration procedures even more than the average man does. People in the Paul camp told me the money spent was invaluable, that it simply would have been otherwise impossible to get Paul doing as many early campaign appearances as he did. Paul was far more on-the-ground in both Iowa and New Hampshire in 2011 than 2007.

I visited Paul's Iowa campaign headquarters in Ankeny unannounced, in early August. Ankeny is a quick shot up a freeway from Des Moines. The office was located near a typical mid-sized Iowa town commercial corner, in a strip mall next to a shorter strip mall. The area around the office is a completely non-epic but in its own way sweeping panorama of American commerce that Paul's free-market philosophy valorizes: a Medicap pharmacy, complete with drive-through for your drugging convenience; a Cazador Mexican restaurant (the border, so far away!); storefronts dedicated to fripperies such as weight loss and nail care that help expand our pursuit of happiness beyond the struggle for food and shelter; a U.S. Cellular store—the web of communications entangling the world, with nothing physical to connect them at all. Ron Paul was born into a world of party lines, if there were lines at all in the home, connecting, by the physical bulk of copper and plastic stretched across lines of wooden poles, the stuff of the earth, the sweat of humans, expended profligately to grant us this gift. Now it all just flows through the air. Paul has a line he likes to use in campaign speeches about cell phones, how there's no way we'd have as many as we have if there had been just a government program, and not a market, dedicated to providing them. These are miracles. These are the sorts of miracles that according to Paul's brand of economics and politics arise from allowing people to innovate, compete, to possess and use property, by spontaneous order and without central command.

Communication, food, self-improvement, commerce— *civilization*. It's easy for those who take the benefits of modernity for granted to mock all that, or deem it less important than it is. It is not martial, it is not grand, it does not redound in any overbearing way to "national glory." But it is the stuff of life as most Americans live it, comfortable and fed and able to seek self-improvement with the excess wealth generated by a culture that roughly follows Ron Paul's prescriptions—private property, freedom of choice in how we use our wealth and capacities. Paul doesn't put it this way, but he is proudly the candidate of the bourgeoisie and burghers, petit and grand—those who make their living not necessarily from hewing the earth, not from commanding men and armies, but by using their property and ingenuity to serve others in the marketplace.

In his campaign office, more than a dozen young people were at their desks working phones and computers. Boxes of campaign literature, Slim Jims (thin, long one-sheet promo material), and schwag filled most of the empty space, including boxes of "liberty fans"—paper hand fans with a Ron Paul branding that read, "I'm a liberty fan." A young woman intercepted me and gingerly asked me my business; I had been having trouble getting callbacks from the media end of the Paul operation and had chosen to just march in unannounced. She summoned Dimitri Kesari from a side office to talk to me. He's the deputy campaign manager from the national office, shipped in to help supervise the Iowa operations in the buildup to the Ames Straw Poll. Kesari asked me to come back in an hour for an interview.

When I returned, the bow-tied and nattily bald Kesari and I retreated to the back room, away from all the workers, along with Trygve Olson, one of the campaign's hot new GOP operative hires, who had just returned from Ames. Olson had more of a dude vibe in comparison to Kesari's dandy; he was rough-hewn and scrappy-looking with sandals on his feet and fast food take-out in his hands.

Most people who were involved in the Paul campaign in 2008 agree that it showed a lack of experience in the nitty-gritty of real Republican Party presidential work, which some interpreted as a lack of seriousness, this refusal to dip into the existing talent pool of national GOP politics. Things were different this time; Kesari was a veteran of campaign work with the Right-to-Work Committee. Olson had gotten his feet wet in the gritty realities of campaigns, winning primaries and delegates, not so much the "air war" game that most grassroots activists love (the wholesale spreading of the name and message of the candidate). Olson worked with Wisconsin governor Tommy Thompson and even—briefly, he stresses—for the McCain campaign in 2008 in its early stages.

Kesari, Olson, and I talked for an hour as young volunteers cycled through the back room. One eager grassroots activist not part of the official campaign volunteer team walked quickly through and handed each of us a copy of the *For Liberty* DVD. (We'll meet that activist again.) Kesari and Olson are both unfamiliar with the DVD, which is a stirring, well-done documentary of the 2007–2008 campaign. I reassured them it's nothing that they needed to be worried about.

Iowa's caucus sets the narrative for the presidential campaign as it rolls from there, though it's also known to set that narrative in a misleading way, with past victories for George Bush in 1979, Pat Robertson in 1987, and Mitt Romney in 2007. But Iowa and New Hampshire are where the Paul people knew they needed to get off to a shooting start, and the Ames Straw Poll is the first publicly accountable way to shine.

Because of Iowa's primacy, Paul had had a functioning office here since May—the Iowa office was in motion even before the national one in Virginia was. Unlike some of the other candidates, Paul already had an enthusiastic grassroots base in Iowa and a handful of well-placed state Republican Party officials on his team. The campaign's ceremonial chair in Iowa was Drew Ivers, a populist-right all-star who had helmed past Iowa efforts for Pat Robertson in 1988 and Pat Buchanan in 1992. While Paul's libertarianism varies from those men's beliefs, an undeniable sociological connection links the three—they are all the candidates of the insurgent antiestablishment Republican, though Ivers might be among the only Robertson men so sure that Paul is also their man. Ivers very firmly told me that Ron Paul is the apotheosis of the freedom-and-tradition message he saw in Reagan, Robertson, and Buchanan.

They had already gotten Paul on the ground in Iowa more often than last time around: 108 events in twenty-eight days through the end of October. They were starting with a good list of supporters already; they had a well-manned phone operation identifying other potential supporters up and running in early August.

The most important new thing they had, Kesari said, was four years of American political history in which Ron Paul's ideas had been vindicated, and seemed so correct to Republicans that his higher-profile rivals were copping them. "Four years ago people were not obsessed with the economy and the dollar; what Ron was saying back then, people were just thinking, 'Oh, he's crazy.' But now they are all echoing Ron, all essentially saying, 'I'm Ron Paul! I agree with Ron Paul!' Lots of people are noticing, wow, he was right on the economy. Even a lot of the blue-hairs, the starched Republicans are going, 'I like this guy. He's hitting the right points. He knows what he's talking about.' Our base has expanded into the Republican base." This allows voters to meet Ron Paul again for the first time—as the prescient spending and inflation and deficit hawk.

It isn't just among potential voters that Ron Paul seems like a more normal Republican. Olson admitted that "if you approached me four years ago and said in the next election you'll be a senior advisor to the Paul campaign, I'd say, that's probably not my politics." But Olson was a hired gun for the National Republican Senatorial Committee in 2010 and was offered the chance to either work on Rand Paul's campaign in Kentucky, or to help unseat Harry Reid in the Sharron Angle campaign in Nevada. "I took a day to think about it and spent some time watching YouTube and reading [about the Pauls]. I had just become a parent, and the issue that was most important to me was, are we going to get control of and get rid of the national debt? I'm the real establishment guy in this [Ron Paul] world, but on the issues I think are most important, debt and deficit, Ron is the only one offering a real solution.

"When I broached the idea of working on Ron Paul's campaign to other high-level Republican operatives, with only one exception they all said, 'That's good; you should do it.' I think there's a real recognition with Republicans that Paul wasn't really treated very well at the convention last time, and that the libertarian constituency is an important part of the limited-government, conservative Republican Party, which has to be a winning coalition." Olson does lament an aspect of Paul supporters that turns off GOP lifers: their belief that they are "two hundred percent with Ron on everything, and they can't be with anyone else. That's not great for party building."

"Conventional wisdom is, you may agree with him, but he doesn't have a chance, so . . . ," Kesari said. "It's like the Goldwater thing: in your heart you know he's right. I often suggest, and people roll their eyes at me, we should print a bunch of Ron Paul T-shirts that say 'In Your Heart You Know He's Right.' A lot of Republicans feel that way." A strong Ames showing could convince them that he's not only right, he's a potential winner.

Paul's Republican opponents were actually noticing his existence this time. Olson had buddies in the Romney camp who told him they were pissed that Paul's campaign hooked an early moneybomb off a swipe at Romney. "And we're going to use part of that million to take a piece out of your guy," Olson told them.

One thing that unnerves the other Republicans about Paul is "they can't take Paul's fundraising base." It's a usual thing with powerful Republican establishment candidates to intimidate any big money that backs a rebel. The Kentucky GOP establishment wanted to try that against Rand Paul and failed, and nothing similar could work against Ron Paul, either. There is no choke point, no carrot or stick that his foes could use to encourage his money to dry up. It's coming from everywhere, in amounts too small and decentralized to intercept. (By the end of third quarter 2011, the Paul campaign had pulled $17 million, from 120,000 donors, with a median contribution of $110.) As Kesari said, "Moneybombs only work for two people, and their last names are Paul."

Digital technologies per se are not the key. Those have to dovetail with a message that resonates with the people using the technologies. Olson remembers a meeting of the International Democratic Union, an international organization of center-right parties, in which an aide to German chancellor Angela Merkel began skylarking about how if he could just get the chancellor's office tweeting, why then . . .

"They didn't get it at all," Olson said. "All Obama's tweeting just reinforced the existing central narrative of young, hip, and connected. McCain could have sent out ten thousand times the tweets and had fifty thousand more Facebook friends and it wouldn't have worked." Ron has the decentralized netcentric fan base he has because he's the candidate of decentralized decision making, liberty, drug legalization, and ending wars. Ron Paul doesn't use the Internet; Ron Paul's *fans* use the Internet, and that is what has mattered.

The big news that week leading up the Ames Straw Poll was Standard and Poor's downgrading of America's bond rating from AAA to AA. While obviously not something to celebrate, as a journalist writing about Ron Paul I took a dark delight in it—the signs of Paul's apocalyptic vision of America's economic future because of the persistent mistakes of its economic past are becoming so clear that only the willingly obdurate could ignore them.

However, the candidate and the campaign didn't seem to see the bright lining (for them) along the gloom descending on the nation. Neither Ron nor Rand Paul nor anyone introducing them at any of the campaign stops that week alluded to it, and these two strategists stopped dead and looked at me as if I'd made some bad joke about someone's recently dead mother when I brought up its potential benefits for them. Maybe this makes them better people than I am, or maybe it makes them politically less bold than they ought to be. But when it came to the debt downgrade, the man and the campaign who told us so was not going to tell us he had told us so.

R on Paul spoke that night in Fairfield, Iowa, a small town with a classic square, green and perfect in the cool golden-hour light. Statues of old men giving sage advice, apparently, to young ones sit near benches, the square circled by old brick store-fronts. Paul appeared in the last hour of mellow dying daylight, underneath a white cupola. Between two trees, a spray-painted Ron Paul rEVOLution banner hung. Four hundred or so people were there to hear him. Fairfield is most famous for being the home of the Maharishi University of Management and a huge Transcendental Meditation center. It's in Jefferson County, the only county in Iowa in which Ron Paul actually won the popular caucus vote in 2008.

A Ron Paul activist from that county explained in a post on

the Ron Paul Forums website how they did it. He credited a personal Ron Paul appearance, a great meetup, a list of 150 local supporters, booths at local farmers' markets and art walks; an office on the town square giving away Paul material; hitting church lots with Paul material; "a focus on local areas of concern, particularly health freedom, restoring civil liberties, opposition to the Patriot Act, restoring the Constitutional limited Federal government, a humble, noninterventionist foreign policy, etc. . . . We directed our efforts mostly to our friends, who were largely Democrats, Independents, or nonpolitical folks. We tried to win over Obama and Edwards supporters, in fact some of our ads contrasted Ron Paul with them. We made a strong effort to get people to register Republican just for voting for Ron." He stressed that they did not do much outreach to traditional conservatives or call Republican voters lists, and "we did not focus on Ron Paul's positions on abortion or immigration [unless asked about them]."

Four years later in Fairfield Paul had attracted families, gray-ponytailed bikers, rockabilly couples, pairs of grandmothers and grandchildren, young dreadlocked men, eager teen girls. Ron strolled onstage, his two black-suited bodyguards nearby. Next to me was an older German man who wanted to talk about money. He agrees with Paul about killing the Fed, but isn't so sure that gold is the proper solution. All money, after all, is fiat when you get right down to it. Curiously, while he granted that Ron Paul is an admirable man of integrity, he also said Paul is "not a Christian." While this man was aware of Paul in 2007, he voted for Huckabee.

As Paul waited to be introduced by Drew Ivers, he sat in his blue-and-white-striped button-down and his tan khaki trousers, alone, hunched over, his hands curled together. Ivers introduced Paul with his usual one-liner—"Let's promote Dr. No to Dr. Veto!" Only Paul, he said, has the expertise and integrity to turn America around and "stop stupid spending!" Paul got up and stood straight and solid at the lectern as he spoke, hands delicately poised, fingers pointing down.

How many of us were here last time he spoke in Fairfield, in 2007, he asked? About a sixth of the crowd cheered. It isn't as if stragglers to the Ron Paul train missed anything then that they won't get now. He's been delivering pretty much the same message the same way for thirty-five years, he said. The difference now is that the country is coming our way. He says things like that—the friendly, happy presumption that the people there to listen share his devotion to liberty. Paul can sometimes, especially on national TV in an unfriendly environment, seem somewhat stern and lecturey and cranky, the grumpy old man of liberty.

On this beautiful day, in front of a happy and enthusiastic crowd, the lightness and cheeriness in the man came out; he was blithe and light and falling back on the light ironies in his style. He tipped his hat to the users and marketers of nutritional supplements, and limned the far reaches of his own radicalism: "Why, I'm so radical I think you have the right to drink raw milk!"

There were some populist nods mixed in—Paul's libertarianism never plays to elite sensibilities, despite the philosophy's reputation as special pleading for the interests of plutocrats. Bankers got the bailouts, he reminded us, while the middle class lost its jobs and homes. And some of those bailed-out banks were *overseas.*

He played the apocalyptician a bit. The official consumer price index (CPI) can't be trusted, nor can official unemployment rates. Things are worse than they are telling us. The middle class is being wiped out before our eyes. And the violence around the world today—the unrest in the European street over fiscal problems—could hit us, too, as the fiat and debt system crumbles. But we need to lose the delusion that somehow war can be good for the economy.

The ideas in Paul's speeches don't connect with sharp clicks. He's fluid, the jazzman sliding in his favorite little motifs that make the crowd go wild. A line that got a standing ovation here in Fairfield: The dilemmas of our overseas interventions are not so very complex: Just come home! Bring all the troops home! As quickly as we can get ships over there to carry them back, bring them home! A slap at the "unwise war against the use of certain drugs" gets about one-fifth of the crowd to its feet.

He's a Republican willing to say that America is no longer the richest or freest country on earth—perhaps the unpatriotic heresy that gets standard-issue GOP voters even angrier at him than talking about blowback. Another revolution is needed—"an intellectual revolution, not a violent one. We have to change minds and attitudes. Teachers and writers must lead the way."

As he loudly proclaimed that freedom requires tolerance, he was nearly drowned out by a massively thrumming motorcycle gang circling the town square. In the kind of gesture that his admirers love so much, he pointed over at the tumbler of water at a table next to the lectern, with a head gesture that said "For me?" Of course it was for him.

He took a tough question from a shaky-voiced, stout older woman in a red polka-dot dress: what about the social insurance programs such as Medicare and Medicaid and Social Security that people rely on? Paul could have avoided the hard question underlying her question, but he slammed right into it: "I believe government insurance of medical care is not constitutional," but instantly ending the programs is not on his agenda. He wants a transition that will continue to take care of those already dependent. But we need to provide a legal way out for those still far from retirement. He did his usual melding of foreign policy and domestic policy: if we stop blowing $700 billion a year on foreign adventurism, that gives us some spare cash to fund this entitlement transition.

As always, a scrum followed Paul as he tried to leave, with wife, Carol, and other family members, to a black minivan parked on the edge of the square. One man with a handheld video camera was trying for an impromptu interview with Paul, even as Paul was already seated in the passenger seat of the minivan. "Do you think Internet gambling should be illegal?" Jeez, man, who do you think you are talking to?

I ran into Eric Boye in the after party of sorts, as dozens of Paul fans and campaign volunteers hung out in the twilight and chatted and ate pizza. I had met him briefly, or at least seen him, earlier in Ankeny. While I was interviewing Kesari and Olson, he had slid quickly through the room and handed all three of us copies of *For Liberty*, then shuffled out. Boye is bearded, earnest, and intense, and enjoys spending his spare time driving around the country promoting Ron Paul. He just drives away from his job running a computer shop in Jersey for a few days in his car with the RONPAUL vanity plates and all the Ron Paul signs, with boxes of Paul DVDs and literature, goes to strange neighborhoods, knocks on doors, and talks Ron Paul. He says most of the people he visits are willing to hang. Interestingly, none of the other people I met in Iowa who do roughly the same thing as Boye—give up their day-to-day lives with no recompense from or coordination with the campaign and travel in the name of Ron—knew him. The movement had grown far too big for everyone to be acquainted with everyone else.

Senator Rand Paul joined his father for a few days of campaigning before the Ames Straw Poll. Someone thought he was a bigger draw than dad—for their joint noon appearance at the Five Sullivan Brothers Convention Center in small downtown Waterloo, the crawling red message on the electronic sign outside read "Rand Paul Iowa Tea Party Tour Today."

Rand, in his blue-checked shirt, tan slacks, no tie, seemed in his Iowa appearances a more restrained and internal man than the scrum of glad-handing politics requires. He's a quiet, thoughtful, hand-to-his-face kind of guy. Then again, he *is* a U.S. senator already. He reminded us of how many federal agencies are out there armed—including the ones on the previous week's raw milk raids in Venice, California. Paul tells various absurd and aggravating tales of government's oppressive leaning on peaceful citizens, including life-destroying fines imposed on a family for the crime of selling rabbits without government permission.

I noticed, with my memories of the man going back twenty-three years, that Ron Paul's face lines have become thick and vivid; still, he moved through his variation of his usual unscripted presentation with agility. He isn't usually big on being bleeding-heart about his antiwar position, but here he alluded to the immense number of men in our military who are suffering post-traumatic stress disorder and committing suicide. He did a little misspeak he makes occasionally, showing how deeply he believes gold is money: about the Fed "just printing the gold."

After hitting gloomy points about the violence in the streets of the European Union countries and the coming collapse of our entitlement state, and how it isn't a burden on our kids we have to fear, it's here on us, now, and how the only people we should be thinking about taking guns from are federal agents attacking innocent citizens, he joked about how he hoped he "didn't scare you off," then asked for our votes in Ames.

Seema Mehta of the *Los Angeles Times* asked Paul if he feels some sense of vindication that so much of what he'd been saying about the economy and money seems so true to so many now. Paul's usual soothing, bobbing drawl laid it out: It's not a matter of personal vindication, he said. He wasn't raised to present himself in any vaunting manner. He won't toot his own horn—and he noted that this bothers his own staff. Saying he won't toot his own horn might seem a meta-clever way *to* toot his own horn, but the lady asked, and he was answering. It's not some rhetorical trick he uses. As far as he is concerned, Paul went on, his real opponent is not Romney or Obama, it's the dead economist Keynes—and the real vindication is for Keynes's economist rivals, the Austrian School, whose teachings Paul is proud to represent.

Rand stepped forward. He's not worried about tooting horns. As far as he's concerned, Rand said, the campaign's slogan should be "Ron Paul Was Right!" As he said this, his father stood with his head down, but smiling.

Paul spent a good half hour in a receiving line afterward. Even with a huge line he'll take a couple of minutes to go eye to eye, and even with people trying to be contentious, as he did here with a young man who didn't like his stance on ending aid to Israel (and everyone else); Paul asked a woman with a couple of preteen girls if the young ladies don't get bored at these political speeches—"Not when it's you!" he was told. He beamed. A young man with a "legalize the Constitution" shirt was greeted as "another radical!" A young man who works at the local Salvation Army, stocky, multi-tattooed, huge tube earrings filling out his lower ear, told Paul he is "the first guy I can vote *for*, not just against." There was a three-generation team of Paul fans in line: mom, daughter, and grandmom.

Later that afternoon, the Pauls were at a hotel in Cedar Rapids, speaking to an older crowd of around one hundred people, filling a room in a Clarion hotel on a typical American avenue of sprawl and commerce. You know instantly when watching Paul speak that he isn't teleprompted or reading a written speech; when you see him speak multiple times, that's not only obvious but a little confusing. Is he a genius reading the pulse of the room when he decides what points he'll make, and in what order? Every talk is just like listening to Ron tell you a random set of things he believes about liberty, government, and some of the things that are wrong with America these days.

Sometimes he sounds like a perfectly respectable Republican candidate for president in the early twenty-first century. Sometimes he chooses to make it clear that there's something much deeper, more interesting, wilder going on. Paul, in his usual un-flashy way, was on fire in Cedar Rapids as a libertarian radical. He stressed a truth about government that libertarians love to use to wow or quiet their statist opponents: that all government power is at root both authoritarian and violent. If you don't do what they order, you'll be arrested. If you resist that, you will be killed. He went off on one of his cultural riffs that feels more progressive left than midwestern-right, about how practitioners of holistic or alternative health care are harassed and oppressed, their vitamins regulated unfairly, their raw milk operations raided. He also lamented the case of Bernard von NotHaus, the marketer of a gold round who was prosecuted for alleged counterfeiting, although the coin does not purport to be U.S. currency. He hit on one of the favorite arguments of the deep-down Woodrow Wilson hater, who recognizes Wilson as the genesis of pretty much all the evils of the modern state, and slams not the Sixteenth Amendment (I mean, Ron Paul having to say he's against the amendment that gave us the income tax would be no more surprising than Billy Graham saying he's against Satan), but the *Seventeenth*. Not many people of any political stripe think about the Seventeenth Amendment, but it upset one of the cornerstones of our federal republic by making senators not the representatives of state governments-as-states, but enshrined the situation where even the second house of Congress will be filled via popular election. It was not mentioned that there is probably no way a Rand Paul could have gotten to the Senate except by the will of the people, but doubtless the principled Paul would have to admit that that would be acceptable collateral damage in the name of federalist principle.

Far from assuring the people how great and grand America is, in the modern GOP tradition, or stressing only its endless promise, in the modern Democratic one, Paul bluntly told that crowd that "we are a poor nation." He was mostly abstract, principled, or historical, not rooted in the specific political controversies of the day, but he did tip his hat to the Paul Ryan plan for spending cuts, while soberly warning that they are not enough, and more soberly noting that the political fire aimed at Ryan even for his woefully inadequate plan showed an American political and media class completely unwilling to face the actual challenges of a bankrupt empire. Yes, the bloody practice of empire can create flashes of fake prosperity, but "war profiteering is not the same thing as a productive society."

Paul is willing to make sure that many of his possible allies also get their hackles ruffled. He is not, he insists, merely saying "just leave me alone, don't tax me, and all will be well," but he does believe that a free society is likely to have more prosperity and a more fair distribution of wealth.

A reporter from NBC asked Paul about that week's stock market drop of more than 600 points. He wouldn't make triumphal hay out of it; for all the apocalypticism built into his message, Paul takes no joy in being a successful Cassandra. Certainly we are now seeing the markets catch up with past mistakes, he said, but we can restore market confidence with a strong dollar. If we could only get a strong dollar. While he didn't spell it out, you remember: End the Fed!

ABC wanted to know: ending the Fed is all well and good, Ron Paul, but what can you do in *day one* of your presidency to bring back jobs? He won't fall into that trap. He understands the impossible complexity of what Austrian guru Hayek called the "catallaxy"—the maddeningly intertwined system of millions of decisions and interactions that dumber politicians and economists reify as the "economy" that the president can influence as if it's a car he's at the wheel and pedal and brakes of. Paul understands it too much to confidently assert that anything he could do as president in day one or day one hundred will have some certain and predictable positive effect. But he had to answer. We can figure that a strong currency will draw in capital, that if we end the tax on dollars coming back into the United States from businesses abroad we can expect more dollars to come back to the States, and if we declare a moratorium on new regulations, we can at least change overnight the business psychology of the country, and that will doubtless be a good thing.

Then, the question for Rand: will he run for president in 2016? Do they think he's crazy? One Paul at a time: this year it's his father.

In Des Moines a few hours later, Larry Pratt of the Gun Owners of America—the gun rights group for those who think the National Rifle Association is a bunch of unreliable sellouts—introduced Paul and gave him a plaque for his service in the cause of Second Amendment rights. Ron looked at the plaque while Pratt delivered the sort of right-wing red meat that Paul almost never does himself, attacking the "liberals" for not wanting to do anything for themselves, alleging that the 1968 Gun Control Act was directly inspired by Nazis. Paul sat and listened and wiped the plaque on his pants to shine it up.

This appearance in Des Moines in the downtown Marriott ballroom was Paul's fourth event of the day, finally after working hours, and featured the biggest crowd at around two hundred. In the afterglow where fans mingle and chat, I met one of the Iowa state legislators supported by Paul's LibertyPac, Glen Massie. He considers himself a Ron Paul type. But he's a vivid exemplar of the fact that Ron Paul contains multitudes, attracting different people for different reasons.

Nothing unlibertarian came out of Massie's mouth, but culturally, in his vibe, in the set of things that concern him, he's of the right-populist-constitutionalist end of Paul, not the explicitly libertarian end.

Massie got very serious with me as we mingled—most of the crowd having dispersed—in the back of the ballroom. He's a former marine who has made his living for years as a diesel mechanic. He showed me his strong, worn hands to prove it. I used the word *politician* when I explained why I wanted to speak to him, and he took quick, joshing umbrage—please don't insult me that way! But then he was quickly serious again as he explained why, in the name of his children and family, he had to enter this ugly world of politics. (He's not the only Paulite public officeholder who insisted to me that I call them "statesman," not politician.)

He couldn't stand by and watch, Massie said, as his nation's constitution was shredded. His list of concerns was culturally deep-right: Judge Rudy Ray Moore being ordered to take down the Ten Commandments from his Alabama courtroom? *Rrrripp!* to the First Amendment. Terry Schiavo, being ordered to die by a federal judge? *Rrrrippp* to the Fourteenth Amendment. Citizens in New Orleans in the aftermath of Hurricane Katrina having their weapons taken from them by government agents? *Rrrrippp* to the Second Amendment. Not on his watch. He was obligated to participate in what he calls the "peaceful revolution"—this whole movement that has been energized by and has arisen around Ron Paul.

Ron was not doing a lot of coalition building himself in his party, but he did get some support from other Republicans here in Iowa. Iowa congressman Steve King appeared in Ames with Paul the morning of the debate, two days before the Ames straw poll. Iowa is, as far as these things go, Ron Paul country in the GOP. Three different county-level Republican Party chairs are Paul fans and were in the crowd today.

Each Paul ended up sitting between me and King as the other Paul spoke. The two congressmen commiserated over some bill of King's that Paul has signed on to; they have mastered the art of conversing so that even the person inches away can't hear. Ron Paul's knuckles were huge and strong.

Rand Paul is better than his father with both the right-wing red meat and some political hokum; with the Paul family bus now here in Iowa with nearly thirty Paul family members from three generations, Rand kept publicly challenging the Romney family to a softball game. (One of the Pauls, Rand says, was a former baseball pro.) The family rode for seventeen hours on the bus from Texas.

When Ron talked about the family bus, he brought up his military record in a way that wasted its red-meat GOP possibilities. He reminisced about a time when he got on a bus and had a very bad time—the bus that took him from Biloxi to Kelly Air Force Base. He never pesters his family on politics, he said. They arrived here of their own free will and on their own initiative, without being asked. Last time he was in Iowa, he remembered out loud, his wife was ill. He knows that Iowans love to meet their politicians, and so he was here more often than he had been in 2007. He reminded us, this being the aftermath of the debt ceiling debate, that the United States has essentially defaulted before. What else can one call the departures from gold in 1933 and 1971, where we refused to pay back debts in the same sort of money that it was borrowed in? That's a default, as well as a radical change in our monetary system—so why act as if a default or a switch back to gold currency would be the end of the world?

The Paul family bus crew skewed young, a real potential dynasty on the hoof. The only one I got to talk to at any length, Rand's son Will, just on the cusp of starting college, wasn't sure a political career is for him. Late one night in a hotel lobby, after the Ames Straw Poll results, he talked about his interest in hip-hop. He doesn't like to write rhymes down—just likes to freestyle, to "spit," in the hip-hop lingo. Spitting—that's what his dad and granddad do when they get up to talk politics, he mused; not delivering a polished professional spiel by rote, but just laying it out from the heart.

The candidate was out of the public eye the rest of the afternoon on Ames debate day. The Family Bus Tour, with Rand in tow, went to Altoona for a pizza parlor presentation, in the back room of the Pizza Ranch. There were about thirty people already there, including Paulian state representatives Kim Pearson and Glen Massie (who was in mufti, in a "Rally for the Republic" T-shirt and shorts). When the family bus arrived, they doubled our number. Pearson, exuding that great, gritty Tea Party mom pep, introduced Rand Paul.

Pearson stressed that she's just your average homeschooling mom, raw milk fan, Tea Party lady, and is beginning to suffer from a lingering paranoia that they might be coming to get her. She waved a jar of raw milk. It was not the first time I'd heard about raw milk during this week of dogging Ron Paul's trail around Iowa. Everyone was alarmed that a few days earlier armed federal agents raided a Venice, California, health food store with guns drawn and arrested three people for selling unpasteurized milk, to knowing and willing customers. In fact, of the two news stories breaking that week—the raw milk raid and the U.S. government's debt rating downgrade—raw milk got more energy and mention from Paul and the people around him. Call it the Raw Milk Revolution.

That he and his people were more inclined to talk raw milk than debt says something important about Paul's mentality, and his appeal. It is not abstract. It feels and is specific and down home, worried about the places where the overwhelmingly absurd hand of government is actually smacking down real citizens in a direct, visible way. Debt downgrades are an abstraction, and exactly what Paul would have expected to happen given the spending and borrowing patterns of the past. But he's not one for I-told-you-sos. He is one for pointing out and decrying injustices and tyrannies that too many Americans—in government and among the citizenry—take for granted or don't notice at all.

I chatted with the two people next to me in the pizza parlor back room: freelance traveling grassroots Paul supporter Britton Sprouse, and Iowan Monte Baugher, a veteran who was stationed in Germany in the mid-1950s. The war was Baugher's issue, and why he loved Paul. Although Paul never spoke for more than ten minutes without mentioning the dangers and excesses of American foreign policy, in Baugher's estimation he still didn't emphasize it enough. It should be *the* issue for Paul, the main thing he runs on, Baugher insisted.

Baugher has a local gang of fellow old men and vets whom he meets for breakfast and a walk every morning. All of his cronies know that the wars are the issue, and that the Republicans and the Democrats are united in being wrong on it. "Every old GI in the VA hospital says: get out!"

Rand Paul is more a standard right-winger in attitude and presentation, if not necessarily in core beliefs, than his father. He'll insult radical environmentalists. He's more apt to slam the entitlement state than Ron is. He's more apt to make it sound like he thinks the rich already pay enough taxes. (Ron is antitax, to be sure, but never says anything these days that even hints at class war, class animosity, or even class bad feelings.) Rand is, though, as apt as his father to sound disconcerting and scary about the looming crisis and day of reckoning that America now faces. "Time is running short." While hitting the reasons we should get out of Afghanistan, he played on a bit of antiforeigner resentment that isn't typical for dad: those untrustworthy Afghans are stealing our soldiers' property in the field!

Ron's wife, Carol, stood up at one point to make something clear—the only time I witnessed her inject anything public about policy, and another example of an important thing to remember when contemplating the future of the Ron Paul movement: no one else is Ron Paul, not even Mrs. Paul. Carol wanted everyone in the room to understand that when Ron talks that strange-sounding talk about the military-industrial complex,

he's *not* talking about cutting defense. Why, Ron Paul is *for* a
strong defense. He's talking about things like, you know, these
huge overseas embassies. Ron Paul means a lot more than that, of
course—Carol doesn't use the word *empire*—but that's certainly
a nice thing for an imagined modal Iowa GOP primary voter, or
straw poll voter, to believe.

Sprouse and I retreated to a nearby Jimmy Johns sandwich
shop; Sprouse had heard they were doing a dollar sub special,
which is gold to impecunious, peripatetic Ron Paul activists.
Alas, the two of us—and Gage Skidmore, a teenage photographer
who had been following Ron Paul around and shooting him for
years for the hell of it and for his fascination with Paul's message
of liberty—missed it by minutes. We ate lunch and talked there
anyway.

Sprouse is a twenty-four-year-old Ron Paul van nomad, traveling the country to support his man; he knows quite a bit about
good places to park and sleep where you aren't apt to be bothered. Even though the campaign explicitly told out-of-staters to
not come to Iowa—too much backlash about the overenthusiastic outside agitators from last time—Sprouse hadn't heard and
showed up anyway, and ran a lot of information tables for Paul
at lots of Iowa fairs. He hears a lot of "he's unelectable" and gets
asked a lot what denomination Paul is. He never sees anyone doing this kind of work for any other candidate; the devotion of the
Paul fan is unique in Republican politics.

I came back from the bathroom and Sprouse had gotten the
entire staff of the Jimmy Johns—around ten of them, recuperating from the one-dollar hoagie rush we just missed—talking Ron
Paul. I had a few of these moments of befuddlement in researching this book, as someone who saw Paul as his indie secret for so
long but now is suddenly confronted with what a big deal he has
become.

They all know of Ron Paul, most claimed he's their man, they mused out loud about whether they'd be able to get off work long enough to get to Ames and participate in the straw poll on Saturday. The whole thing felt like Sprouse set it up as a prank during my forty-five seconds away. But no, everywhere you go, people were eager to talk Ron Paul. They didn't necessarily have a deep understanding of the whole Ron Paul picture. But they knew, a little, and they liked, a lot. "He makes everyone else look like idiots," said one sandwich maker.

I watched the Ames debate that night from the press room, segregated from where the candidates were speaking. From tables lined up cafeteria-style, I and the rest of the press corps watched the debate on jumbotron screens. (Neither I nor my seatmate from an Iowa newspaper, also doing his first debate coverage, realized that press passes don't actually get you in the room where the debate occurs. Take that, mainstream media!)

This Ames debate featured one of those classic "only from Ron Paul" moments that make some of his fans adore him all the more, some of his fans worry that he's shooting himself in the foot, and his enemies believe that he's a dangerous lunatic. The question of Iranian nukes came up. And why, Paul wondered on national TV from the stage of a Republican presidential debate, shouldn't Iran want, and by implication have, a nuclear bomb?

"Just think of the agitation and the worrying of a country that might get a nuclear weapon someday," Paul said. "And just think of how many nuclear weapons surround Iran. The Chinese are there. The Indians are there. The Pakistanis are there. The Israelis are there. The United States is there. All these countries—China has nuclear weapons. Why wouldn't it be natural that they might want a weapon? There'd be—internationally, they'd be given more respect. . . . You know, in the fifties, we at least talked to them. At least our leaders and Reagan talked to the Soviets. What's so terribly bad about this? And people, countries that you put sanctions on, you are more likely to fight them. I say a policy of peace is free trade. Stay out of their internal business. Don't get involved in these wars. And just bring our troops home."

Whew. "The United States is there?" Heavy stuff for a Republican trying to win votes—reports from Paul workers indicate they knew for a fact it cost him a couple of hundred votes at the Ames Straw Poll, which mattered.

It was like the "legalize heroin" moment all over again, foreign policy division: an absolutely clear and commonsense declaration so in opposition to conventional wisdom one scarcely knew what to even say about it.

After the debate, the one advantage of the press pass came into play: our auditorium floor became the "spin room" to which candidates' handlers and spokespersons and even some candidates were sent to chat with reporters, explain what they said and why it meant they won, and get interviewed by Sean Hannity live on Fox, one of the debates' sponsors. Rick Santorum was still unable to stop talking about how unbelievable it was that Ron said that—well, Santorum added with a jokey knowingness, he should have *known* Ron Paul would say something like that, but still—unbelievable! Jesse Benton, the campaign's political director, and Rand were sent in to speak on Ron's behalf. Neither was very comfortable with the Iran comment, or willing to endorse it in its obvious radicalness. They stumbled around about what he really meant to say, and how we managed to have détente with a nuclear Soviet Union, but certainly Senator Paul did not want to be on record saying he doesn't care if Iran has nukes.

The day after the debate, the Iowa State Fair. Yes, as you've likely heard, it is full of butter sculptures and fried food, and lots of midwesterners trying to have a good time together, and it made for a perfectly pleasant day. Ron started his day with a live appearance on the Jan Mickelson show on WHO-AM, broadcasting from the fair in a glass-walled room. I and a few other fans and curious passersby lurked on the outside, looking in. One chunky

African-American man sidled up next to me and recognized Ron, and told me there's a lot he likes about Paul, particularly regarding the wars. But he thinks Paul is "weak on protecting the little people." In the dream laissez-faire world of Ron Paul, he said, kids will be stuck working in coal mines and the elderly will be sunk in poverty. Sure, he sees the benefits of balanced budgets—but you can't balance budgets on the back of the poor. Do citizens sound like TV pundits now, or are TV pundits trying to sound like citizens? The second person who asked me why I was lurking outside frantically taking notes as we listened to Paul's piped broadcast was a Ron Paul road warrior who had trekked in from Chicago for the day just to see Ron. Another passerby mentioned being disturbed that Ron Paul was for earmarks (targeted spending proposals stuck into bills by congressmen), which had become a bête noire of many right-wing anti-spenders. Some of Paul's foes slam him for his unusual take on the earmark question. Paul has a record of proposing earmarks that aid his district, then voting against the overall spending bills that include them, a move that strikes some as hypocritical. Paul defends earmarks philosophically, as keeping a tighter control on where government money goes in the hands of Congress. Congress, after all, is the branch closest to the people, responsible for the power of the purse, and Paul thinks earmarks help keep that power close to Congress, rather than giving the executive its head to distribute federal money as it wishes.

For a notorious peacenik and magnet for peaceniks, Ron Paul sounds often like a penny-pinching Republican politician when discussing the war issue. Especially before GOP primary voter prospects, he is far quicker to emphasize the crushing costs of empire in terms of cash than in terms of lives. Mickelson told me back in 2007 that your standard Iowa "values voter" found Paul's "war policies confusing and irritating. They don't understand how you can be a constitutionalist for limited government and be against the war and not be aiding and abetting both al-Qaeda and MoveOn.org." Talking with him now, Paul used the unambiguous word *tragedy* to refer to what's going on overseas as we conduct our wars.

Mickelson, who has talked to them all, also told me what was unique about Paul as a politician and interview subject: "If you talk to him, he answers the question you asked him. That makes him a very good interview but a crummy politician. Most realize you can't answer questions and be elected. You can't actually engage."

Mickelson pointed out something only a Paul fan would have noted about the previous night's debate: when it comes to banking and money issues, now everyone was Ron Paul. Gingrich seemed tougher on the Fed than Paul himself. Mickelson spelled out some of the scary news inherent in the Paul message that Paul himself usually doesn't lean on too hard: that any way out of the current economic mess, any "correction," is unlikely to be a soft landing. Ron didn't disagree, exactly, though he pointed out that it doesn't do anyone any good if we pay off the face value of our Social Security obligations, say, in grossly devalued paper currency.

Rather than referring to China as either a potential military enemy or an unfair trade partner, as is typical for his GOP colleagues, Paul talked of how China is busy buying assets and trading with the world, not waging war and blowing up assets. Paul scoffed at the notion that official intelligence briefings he might have access to as a congressman might be useful—he doesn't even bother with them, he said; they are just propaganda. He learns more about what's really going on in the world on the Internet. Paul mentioned his colleague Walter Jones, Republican from North Carolina, who later regretted his support of the Iraq War, support based on lies he gleaned from such briefings, as he watched American soldiers come back from the war wounded, damaged, or not coming back at all.

During the break, Mickelson, whom I'd interviewed before about Paul and Iowans, gestured me into the booth with Paul, Drew Ivers, and A. J. Spiker, a former Iowa county GOP committee chair on the Paul team. Mickelson explained that lots of Paul fans around the world were listening live via the Internet. Paul was intrigued—how do they know how and when to listen? He is fascinated but unsure of the mechanics of how his international circle of fans communicates about him.

Does he not know thousands of people spend all day discussing him, charting his moves, dissecting his enemies, his every pronouncement? It seems he does not. Traveling with Paul was described to me by one of his 2008 handlers, youth coordinator Jeff Frazee, as being like traveling with the Beatles; Paul has indeed maintained the winning humbleness that his devotees speak of, and one way he does that is by not thinking too hard about what those dense crowds of people mean, the ones he has to force his way through slowly after all his appearances.

He really never thinks it's about him, even when it clearly is. He once wondered to me as a fellow old libertarian movement hand: what happened in the past four years, to create these huge crowds of excitable liberty-minded kids? Well, a lot has surely happened. But the main thing that happened is that *he* happened. Well, hmm, is his response to that idea. He doesn't want to think of it that way.

Mickelson teased Paul about how he was corrupting the youth of Iowa, ruining a generation of GOP activists, infiltrating the Iowa party with alien ideas. Paul did a shuffle like the proverbial little boy caught with his hand in the cookie jar, and looked down: "I feel badly," he said with comic bashfulness.

Back out in the after-scrum of fans and Paul family members, I met the Menkhauses: mother Tasha and her two daughters, Tara, seventeen, and Brynn. Brynn, twelve, was carrying her own homemade, hippie-colorful "Ron Paul is the Constitution" sign. She had already written a nearly four-hundred-page fantasy novel, and wants to write another one dramatizing the dangers of hyperinflation. I recommended she check out libertarian science fiction author J. Neil Schulman's 1979 novel on that theme, *Alongside Night.*

The Menkhauses had spent their whole summer on the road for Ron. On days when they weren't at an event he was at, they'd set up a table and give out literature outside whatever drugstore they could find. One of their missions before they left was to collect some more campaign lit from the Ankeny headquarters, which they did. When I ran into them again and again at all the rest of that week's Paul events and chatted, it was impossible not to get across my feeling that there was something a little weird about this, the mother and her two daughters spending their summer this way. Ms. Menkhaus couldn't really address the issue, because she didn't see anything weird about it at all. She and her daughters are all passionately devoted to Paul's message, and they think it's desperately important the country hears it. Brynn can get genuinely frightened when she thinks of the civil liberties depredations her government currently gets away with and worries even more about what might be up its sleeve next.

I drove that evening to the Pilla Opera House with Christopher David, the principal behind the Paul movement news and commentary website Revolutimes.com, and a candidate to unseat Henry Waxman from his congressional seat in California in 2012. David told me of his attempts—ultimately fruitless—to get Young Americans for Liberty to ally with and promote a "Year of Youth" campaign that would feature and promote actual young *candidates*, not merely train and encourage youngsters to work for other people's campaigns.

David is a great combination of salesman and earnest politico, and represents a big part of the spirit of the Ron Paul world—as both businessman and activist he wants to forge a life that's all liberty movement, all the time. He was on the verge of launching his liberty movement news and analysis portal site Revolutimes, which hopes to grow to a sort of *Huffington Post* for the freedom world. (His ally in this is Richard Williams, the Los Angeles lawyer and Senate candidate.) We were going to Pilla to join dozens of other Paul fans to watch a showing of the great 2008 grassroots campaign documentary *For Liberty*. The night ended up long and raucous and hilarious, as the charming sisters who run the Opera House opened up their well-stocked refrigerators of beer and regaled us with funny, scabrous tales (that they ordered me not to write down) of Sarah Palin's world premiere of her documentary in their lovely restored theater. David sat alone at his laptop, away from the noise, working working working.

The next day was the Ames Straw Poll. As various people on the ground in Iowa reminded me, the straw poll is mostly a fundraiser for the Iowa Republican Party, one designed to milk as much cash from the vying candidates as possible. Votes are, quite literally, paid for. But not directly. Votes do have to be cast for candidates by actual human beings, one ticket or vote per person. They even mark a finger with paint to prevent gaming the system. Just like in those photos from "liberated" Iraq.

The tickets cost thirty dollars, but that doesn't have to be paid for by the voter. It's paid for by the campaign. But since the ticket is merely something that *allows* someone to cast a vote, not a vote itself, it doesn't matter how many tickets the campaigns buy if they can't deliver bodies to the campus of Iowa State University in Ames on a warm summer Saturday, bodies willing to cast those votes.

Most campaigns simply give them away to all comers who they are sure will vote for them. Paul's campaign is a rare one that actually sells them—though at a discount. The campaign was making them available to supporters for ten dollars in the weeks leading up to Ames; as the day drew close, and certainly as the day wore on at Ames itself, they began giving them gratis. As campaign consultant Trygve Olson said, the straw poll proves depth of support, if not breadth: participating asks a lot of the voter, essentially a full Saturday of their time. It's not a matter of just going to a local school and casting a vote, and certainly not as easy as telling a pollster on the phone that you support a certain candidate. Getting thousands of votes says a lot about your organizational savvy and how much those voters love you. And they all have to be Iowans. This is not something Paul's campaign could game with outside agitators. A voter that loves you is a voter who will likely do their best to encourage their family and friends and coworkers to vote for you.

Michele Bachmann was essentially paying people *to* vote for her. She brought in country star Randy Travis to sing in her enclosed, air-conditioned tent. You were only allowed in to enjoy him if they were satisfied you had voted for her—she had her people leading groups of supporters to vote and then bringing them back.

Paul had ponied up the largest amount for the largest footprint on the straw poll grounds—$31,000. Paul's people had told me it was well worth it, true prime real estate. When I arrived poll day at the official opening time of 9 A.M., I couldn't quite see it. The parking lot that I and thousands of others were funneled into had an access path into the grounds of the Hilton Coliseum, around which various tents for the candidates and causes were scattered. That natural walking path from the parking lot went between two buildings and back to another big open space in which Bachmann's tent dominated, with Herman Cain to the left, Rick Santorum to the far right, and sad but persistent mystery candidate Thaddeus McCotter immediately to the right of Bachmann's. Bachmann had golf carts cruising the parking lot, ready to take you in—straight to her tent, of course.

Paul's space was in its own square off to the left of the pathway, and it seemed to me it was easy to avoid or miss entirely if you were inclined to head straight back to Bachmannland. Bachmann had a phalanx of orange-T-shirted youthful volunteers (not a one of whom looked to me like they genuinely wanted to spend their Saturday volunteering to help Michele Bachmann become president).

For the first hour or so, no one in Paul's area was filling the space or stopping any passersby to give them the word of Paul. Most of Paul's suited campaign workers and volunteers were huddled underneath an overhead walkway in back. The kids' play area—the "Prosperity Playground"—rang with no gleeful shouts. The "sliding dollar" inflatable slide sat unused. The stage where various Pauls and musical performers would do their thing later was empty. Bachmann's area was already a scrum, hard to navigate by 9:30. I thought it looked grim for Paul.

Campaign consultant Trygve Olson had told me the straw poll was now telling campaigns they couldn't have their people march around demonstrating their support—you were supposed to keep your energy to your own space. He thought this was deliberately aimed at the Paulistas who had rocked the straw poll in 2007 with a rolling parade of Paul fever. As one Iowan Paul activist told me about 2007, "The Republicans were worried the march was gonna turn violent, that Ron Paul supporters were gonna start rioting because we were so passionate. Every candidate should have people doing that! And it was all volunteers, people who wanted to be there, weren't paid, weren't on staff, there on their own time. We saw other candidates paying people, shoving them in like cattle. I wouldn't even want to support a candidate like that."

By eleven, things were hopping in Paulville, kids playing, hundreds enjoying hot dogs and corn and beans under a huge tent, Jordan Page playing his Ron Paul folk songs to hundreds of listeners ("If freedom's the answer, then what is the question? Resist the forces of force and aggression . . . as God is my witness I'll be a slave no more"), both Pauls and prominent staffers tag-teaming the stage all afternoon. Barry Goldwater Jr. was lurking about, with that pure amazing ripped-from-history Goldwater look, that granite chin, that cowboy timelessness, and a strong twist to the mouth that speaks of righteous disgust, that "you knuckleheads better get your act together" grimace.

I canvassed some voters on line at Cain's tent—yes, Godfather's Pizza was being served. One man was there with his wife; the kids were at home. He's a project manager for small businesses and wasn't sure who he was voting for. He said he knows Paul but doesn't love him—but appreciates that he's a "flamethrower" and that might be just what the party needs. But then again maybe a transpartisan compromiser has to get the country moving again? He's a *National Review* reader, he said, and does worry about electability for the GOP's candidate. He sees and appreciates Paul's steady philosophy and principles but . . . some military disengagement is surely called for, but . . .

For someone trained in libertarian rationalism about political philosophy, the gooey vagueness of the typical Republican undecided is maddening. But I did learn around Ames that Paul can potentially appeal to more than the already-Paul diehards. They aren't all afraid of him anymore. But they are a little doubtful he can play in prime time.

Inside, the Iowa Republican Party showed a video presenting a fair bit of economic resentment—women weeping over lost economic opportunities in the Obama age, a World War II veteran lamenting that his kids are saddled with debt. One hears very little of these personal stories of troubles from Paul fans—their concerns seem animated more by principle, less by personal pique or circumstances.

When Paul gave his speech inside to the assembled folk, not his chosen fan base, he did something I never saw him do before or since: open with abortion. He even drew attention to it, saying that he usually talks about liberty and the Constitution bringing people prosperity but today he wanted to emphasize something different—life. The prime reason government exists is to protect life, all life, and liberty and life are both from the creator. (Despite this, the prominent Iowa-based social conservative group Family Leader in November refused to endorse Paul because he doesn't sufficiently promote federal solutions to abortion or gay marriage for their taste.)

Over lunch, provided by the Paul camp, I chatted with a Transcendental Meditation devotee from Fairfield who views Paul as a new-paradigm politician, and with Tom McIntyre, whose wife, Crystal, was a delegate to the Republican National Convention last time around and found it frustrating—being ordered around, leaned on to wave a McCain sign when she was a Paul woman. She showed up today and forgot her driver's license, which meant she wouldn't be able to vote. A fellow Paul fan offered to drive her home, a round trip of several hours, to get it. McIntyre said his wife is ready to give up on politics if Ron doesn't win—what about Rand, he'll argue? "Ron's message often falls on deaf ears, like the Gospel," he said. "If it's not part of the daily set of ideas people hear, they don't want to know about it."

I chatted with Vincent Campos, an Iraq war vet who voted for Paul in the Iowa caucus last time around, then moved to Texas and voted for him again there. He was a national delegate to the Republican convention—the Texans loved him because he was a young veteran—and voted for Paul, though he has no idea if the head of the state delegation announced or counted the vote officially. He goes around Iowa handing out Paul pamphlets after church services. A Santorum guy upbraided him for politicizing church. He told the Santorum guy he was just jealous because Santorum has no young supporter willing to canvass for him.

Ivers got onstage and told us this is the most important election of our lives—he said he knows we've heard that before, but he wants us to know this time it's true. He told us the campaign has helped organize forty-one different special issue ". . . for Ron Paul" coalitions, including the deaf and Seventh-Day Adventists.

As the votes were being counted, I sat in the coliseum with a bunch of grassroots volunteers, including the Menkhaus family—mom Tasha brought some local pastries for us all—and we handicapped what might happen. There was a lot of strong hope that Ron had it in the bag.

Paul came in a very close second, losing to Bachmann by only 152 votes, less than 1 percent. That was great, in a way, and hurt, in a way. While they didn't and won't say it for the record, the higher-ups in the Paul campaign were pretty sure they had done their advance work well enough that they thought they'd score an actual victory—not merely the close second place that got them almost no media play and temporarily pushed Michele Bachmann into "serious front-runner" status. They made the calls, they distributed the tickets, they rented and ran the dozens of buses. A fried Steve Bierfeldt, who ran Paul's Iowa operation, told me the following Monday that "no one wanted to win more than I did. I haven't had more than four hours of sleep a night in three months."

After the vote, the Paul family and campaign team retreated to the Ames hotel where Paul was staying; word spread pretty well among Paul fans and a couple dozen of us well-wishers and fans went there to hang out as well. Despite the news value of Paul's accomplishments, I was the only journalist there. With no fanfare or entourage, Paul came down to the lobby, where he was congratulated and sympathized with; we chatted for a minute, and it seemed he genuinely had been expecting a win. He had done the math in his head—he knew his people did their best, but if we had just turned seventy-six votes from Bachmann to him . . . but, he sighed, when he gets discouraged, he remembers that they aren't just trying to reverse ObamaCare or a decade of bad decisions; they are trying to reverse a hundred years of bad attitudes toward freedom and to really change history.

Paul told me all this quickly as we stood near the hotel bar. Sitting there drinking already was Goldwater Jr. He leaned back, wearing that iconic face, placed his hand firmly on Paul's shoulder, and asked: "Do you need a tequila shot?"

No, he did not. The night devolved into a cheery party, Paul troubadour Jordan Page turning himself into a human karaoke machine for various younger-generation Paul girls and men singing into the night. Paul walked the room and paid his respects to everyone who wanted it before he retreated; he chatted anarchist philosophy with one (Paul admitted that he does think about the value of private defense at times, and that too many police officers just violate rights rather than protect them, but . . .), defended the RevolutionPAC (a Super PAC of Paul fans explicitly annoyed with the official campaign's messaging) with another (though he lamented that he can't directly call on the wisdom or help of any of its principals because of campaign finance law—"maybe they'll find money we wouldn't have"), and talks with twelve-year-old Brynn Menkhaus about the Fourteenth Amendment, which he thinks gives too much power to the federal government and practically kills the Ninth and Tenth.

I met the Fiscus family, Kevin and Marie and their teen son Steele. They do international mission work. They thought they might be Bachmann people, until they saw her unable and unwilling to answer serious questions in a TV interview. They bought ten tickets for the straw poll but only found seven people to cast them. Various friends told them, well, they sure believed in what Ron Paul had to say but they didn't think he was electable. Marie was a bit angry: "I might need to find new friends, because you should fight for what you believe in."

The debates continued as the fall campaign season progressed, more debates than either the public or candidates knew what to do with; Paul admitted to me he's bored by them, and joked about a fan's suggestion he ostentatiously read a newspaper while on the podium. (That's of course not the sort of thing he'd ever really do.)

The first Republican Party debate in September was held at a place with curious and complicated resonance for Ron Paul: the Ronald Reagan Presidential Library and Museum in Simi Valley, California.

Paul's history with Reagan, and the memory of Reagan, makes an interesting case study in his own evolution as an American political figure. In 1976, the year he entered Congress for the first time, Paul was bucking the dominant trend in the Republican Party by being for Ronald Reagan. Reagan had first bubbled as the great conservative hope after Goldwater at the 1968 Republican convention, when he was the right-wing pole opposed to Nelson Rockefeller's left wing (of the Republican Party), each trying to steal the election from Nixon, and each failing. Nineteen seventy-two was then Nixon's again, and thanks to Watergate, 1976 was Ford's—with Rocky as a running mate. But even in the 1970s, Ron Paul could never be a Ford-Rockefeller man.

Being an old-school Reagan loyalist meant a lot then, and it meant a lot to Paul. Which is why his more-in-sorrow-than-in-anger farewell to the Republican Party in 1987 prior to his Libertarian Party run was so sincerely meant. Reagan had stopped being Reagan. Just as he never wants the loyalty of his own supporters to be about the man rather than the ideas—though it often is—Paul's loyalty was to Reaganism, in the original sense, not Reagan.

Before this debate, in the sacred halls of Reaganism, something happened that showed how different things were for Paul this time around: one of his front-running opponents trouncing him in the polls decided to go for a public attack. Rick Perry issued a press release that reproduced Paul's 1987 disappointment with Reagan.

It's hard to believe Perry's people couldn't anticipate how Paul's camp would react. Paul came back, both in a written statement and live in the debate, with his defense of standing by the meaning of Reagan the limited-government conservative, not religious obeisance to his saintlike memory.

In front of the library—well, not really in front of the building per se, no one who wasn't an invited guest was allowed to get anywhere near it—dozens of Ron Paul fans showed their love to the grandees, press, and Friends of Reagan pulling into the property. No one at all was there to support any other candidate. A handful of union members, disgruntled Democrats, and members of a gay rights group were also doing anti-Republican Party sign waving, but they were lost in a sea of Paulites.

I walked the line to meet as many as I could. Los Angeles Ron Paul fans are, as Paul Stanley of Kiss might say from the stage with swagger, the rowdiest, wildest Ron Paul fans I have *ever* seen. They run the range of types Paul appeals to, including the types that some Paul fans don't like to think about.

I met a motorcycle enthusiast journalist with a long past as a libertarian and Bircher who admitted that his coworkers get annoyed with him for talking too much Ron Paul. (There's a pretty funny video circulating, available on YouTube, one of those computer-generated cartoons, in which a newly minted Paul fan babbles "Ron Paul. End the Fed" and other Paulista catchphrases until his friend suggests she thinks he's having a stroke.) Next down the line was an older gentleman who remembered with a laugh voting for Wallace—not George, but *Henry*. He's learned a lot since then—for instance, that the Birch Society is controlled by *them* and that most of the 9/11 Truth movement are cowards because they don't firmly and loudly call out who was really responsible—Israel. He calls these cowards "9/11 Half-Truthers."

Then there was an aerospace engineer heavily into online Paul organizing. Then a milling machine engineer who likes to stress the economy and Patriot Act when selling Paul to family, friends, and acquaintances. Then a former self-described "GOP Yes man" in an End the Fed shirt who "came to the revolution late," after the last Paul run, after realizing that his own party was guilty of the same offenses he liked to slam Democrats for. A guy told me that his Mexican parents get the hard-money message because they remember the peso becoming worthless.

A Ford minivan pulled up and a bearded guy leaned out. "I love Ron Paul," he told us. "How do I join you guys?" He was given instructions about the nearest legal place to park, but, no, no, not literally join us in sign waving this second, but hook up with Ron Paul activists later. Meetup.com information was shouted at him.

I wouldn't have believed this myself if I hadn't witnessed it: At least every thirty seconds, on this road in Simi Valley, and including from a fair number of cars turning into the debate itself, someone honked their approval of Ron Paul. Spontaneous public expressions of Ronlove are far more frequent than the national polls would lead you to think. A garbage truck honked for Ron Paul. A bus full of senior citizens being driven to their retirement home honked for Ron Paul.

Across the street I met Patricia Winkler, who makes political videos for YouTube. Does she think this sort of public display of Paul love does much to change the world? Sure. It creates community, for one; these people meet, socialize, communicate, give each other strength and encouragement to continue with activism, help each other hone their understanding of the truth and ability to convey it. And she thinks the politicians *do* notice these things—and assume that every one person making a public fuss represents another hundred who think the same thing but aren't on the streets.

That formula may apply to how, say, a congressional office interprets constituent mail. But with Ron Paul, the political powers that be and the media seem to assume the opposite: that Paul and his message have *nothing but* this enthusiastic and small mass who can be counted on to flood Internet polls, wave signs on corners, vote for him at straw polls, even at their own expense—but that they do *not* represent any larger constituency that requires respect. And of course, the standard supposition goes, they are all painfully eccentric, bordering on nuts. It's like a jokey line from an old *Time* feature on Paul from 2007: a staffer for Congressman Tom Tancredo of Colorado "walked up to a guy in a shark costume and asked him if he was a Ron Paul supporter. 'No. They're all nuts,' replied the shark. 'I'm just a guy in a shark suit.'"

I drove about a mile down the road to a spacious pan-Asian restaurant and bar called the Elephant Bar, which was crammed with 150 Paul fans watching the debate on TV screens. There was no looking away from the debate—big screens in every direction. Someone was holding a Ron Paul skateboard, decorated with a great painting of the man with the "truth is treason in the empire of lies" line.

It was Paul's most forceful and smart debate performance yet. He realized that nearly every question was designed to trip him up on some matter of libertarian purism—to force him to cop to some outside extreme of the live-free-or-die philosophy that the interlocutor assumes will ensure no decent American could dream of voting for him—and Paul never got trapped. He laid out a brilliant rhetorical flourish to explain his seemingly abstruse monetary economics, shocking us into a Zen moment of clarity with a seeming absurdity that quickly became an obvious truth. Paul boldly declared that he could get the price of gasoline down to a *dime*.

At that point John Harris of *Politico* tried to cut him off, leaving that seemingly dotty claim the last thing the viewer heard from crazy Uncle Ron. Paul wouldn't let him get away with it, and explained that an old-fashioned *silver* dime would be worth at least $3.50, the price of gas. This is a point about inflation that's hard to grasp intuitively, and the dime example should have helped make it clear. It isn't that the things we need to live, in many cases, are inherently getting any more expensive: when you just compare silver to oil and leave the dollar out of it, you see that there's little difference in the exchange value over decades, on average. It's the dollar that's getting worth less, and that's why we must End the Fed.

Rick Perry at this debate pulled from Paul's radical libertarian playbook, resolutely jumping on the so-called third rail of politics and skateboarding down it. Social Security is not only a mess, not only unsavable as currently operating, but inherently an impossible mistake, a Ponzi scheme, not the sensible and workable investment vehicle the government claims. Its ability to pay out depends on finding a steady stream of new suckers—young taxpayers—to join in. We are all forced to do so, so the line of suckers is endless, so what's the problem? The problem is we're seeing fewer and fewer young taxpaying workers whose incomes can be tapped to make good on the promises to the baby boomers.

It's a longtime libertarian policy wonk argument. The current sense of debt crisis is the only thing that could lead Americans—all of whom are supposed to expect to have their golden years subsidized by government paper—to seriously rethink the program. Yes, that debt crisis that Ron Paul was the first and loudest voice in the Republican Party to note with alarm.

Paul agrees with Perry on the basics of the problem with Social Security. But he doesn't choose to make assaulting it a big part of his platform. For sure, he thinks it's unworkable and unconstitutional. But he's acutely conscious, in a Hayekian way, of all the plans and expectations built around its near century of existence, and he thinks it would be neither wise nor kind to upend them instantly. He thinks promises already made for those near retirement should be honored, while the younger should be given a means to escape the forced fake-savings program. Paying the government's obligations on this and other so-called entitlement programs can only happen with serious cuts elsewhere—and such cuts have to come with a rethinking of government purposes and responsibilities. But merely making a frontal assault on Social Security is not high on his rhetorical agenda.

Then, the immigration question came. It's one of the issues where Paul is slammed by other libertarians for being insufficiently libertarian. While very little in his actual writings after 2007 could not have come from the pen of an anarchist—and some of his anarchist fans are sure Paul is one of them—when asked Paul insists he's not. (He's clearly at least sympathetic to the ideas, as if he can see anarchism as an ideal to be wished for that fallen human nature, alas, cannot allow.) But when challenged from the libertarian side on immigration, as I did when interviewing him at the launch of his 2008 campaign, Paul will fall back on the constitutionalist stance and point out that defending and securing the border is a legitimate function of the federal government.

If a libertarian doesn't see the sense in his immigration position, Paul said to me in January 2007, "they'd have to be anarchists, and I'm not. I believe in national borders and national security. My position is, take away incentives—why are states compelled to give free education and medical care? I don't endorse easy automatic citizenship for people who break the law. They shouldn't be able to come reap the benefits of the welfare state. I don't think libertarians can endorse that. I think removing the incentives is very important, but I don't think you can solve the immigration problem until you deal with the welfare state and the need for labor created by a government that interferes with the market economy. We're short of labor at the same time lots of people are paid not to work. Take away [illegal immigrants'] incentives. I do believe in a responsibility to protect our borders, rather than worrying about the border between North and South Korea or Iraq and Syria, and I think that's a reasonable position."

But he lived up to the highest hopes of the free-immigration libertarians, especially in comparison with his competitors, in how he chose to swing at the issue when asked at this Reagan library debate. He took it to the drug war, which no one else would mention: "And it's the drug war that's going on there. And our drug laws are driving this. So now we're killing thousands and thousands of people. That makes it much more complicated. But the people who want big fences and guns, sure, we can secure the borders—a barbed-wire fence with machine guns, that would do the trick. I don't believe that's what America is all about. I just really don't."

He ends with a line that dances on the borderline between kooky paranoia and visionary, the kind of line that the experts think destroys his reputation but actually makes his fans love him all the more and marks him as someone with a particularly brave form of common sense to many others: "But every time you think about this toughness on the border and ID cards and real I.D.s, think that it's a penalty against the American people, too. I think this fence business is designed and may well be used against us and keep us in. In economic turmoil, the people want to leave with their capital. And there's capital controls and there's people control. So, every time you think of a fence keeping all those bad people out, think about those fences maybe being used against us, keeping us in."

Ron Paul. No one like him.

Paul was doing pretty well in these early stages of the campaign. His second-quarter fundraising numbers had him number two only to front-runner Romney, at $4.5 million. The press narrative still couldn't quite swallow him whole, though; a *Washington Post* analysis of the field on the day those figures were released mentioned that Romney was ahead of everyone, then listed every other candidate and their total except Paul. The new excuse, expressed baldly by the *Post* in a different article, was that it was fine to ignore him last time because his ideas were too outré; it was fine to ignore him now because his ideas had become too common in the GOP field. The *Week* quoted Fran Wendelboe, a former New Hampshire Republican state representative: "The conservatives who might have gone with him in the past have enough other choices this time."

I tagged along with Paul at a mid-September convention thrown by Campaign for Liberty, called LPAC; Paul had Vince Vaughn in his entourage all day. Paul told the luncheon donor crowd that he's always considered his antiwar message *the* core of his campaign; at a small eight-person private donor meeting later in the afternoon he skipped any stump speech, chatted with the donors, and wondered aloud a lot about the strange phenomenon of left-progressives hopping on his bandwagon.

Candidates can get places reporters cannot. In order for me to catch both Paul's evening speech to LPAC in Reno and his morning speech to the state Republican Party convention in Los Angeles the next morning, Paul's political director Jesse Benton let me tag along on the small eight-seat private jet the campaign leased. Four seats were face-to-face, with my legs nearly touching Paul's. Paul's congressional chief of staff, Jeff Deist, was to my right, Benton diagonal.

Paul was going over his notes and schedule for the day. Presidential candidates, seeking the most powerful office on earth, are very little the masters of their own destiny. Some of the most seemingly exalted—high-level politicians, executives, entertainers—often lack ground-level autonomy, their days consisting of going where the people they hire tell them to go, doing what the people they hire tell them to do.

Benton explained to Paul what was going on that day in Los Angeles—a convention of the state GOP—and why they were there: they'd been invited, and a straw poll was happening that it would be good to do well in. He was giving a breakfast talk, appearing before a special panel on medical issues, and then addressing a group of his local grassroots fans, operating under the banner of the Republican Liberty Caucus.

As Paul listened to Benton explain his commitments, his mouth was firm, and he shook his head slightly. All right, he announced with undue firmness after he heard Benton's explanation: "I'll go through with it." (Paul, whose public reputation is all sternly or dottily grandfatherly, has a streak of sly and silly irony that comes out often in smaller crowds.)

This was an informal session. My tape recorder and notepad were not out. But we chatted about Rick Perry. Paul, as is his wont, doesn't get personal with the words out of his own mouth. (This year the campaign did begin targeting his opponents specifically in ads, something that annoys some of his purist fans, though the campaign rightly sees it as playing presidential politics to win, the way it must be played.) I offered my impression— picked up from some sideways comments and expressions I saw while dogging his trail in Iowa the week Perry entered the race— that he had an animus toward his governor more pointed than usual, possibly with a personal edge.

Paul didn't confirm that. He merely offered that, politically, Perry struck him as highly authoritarian, and he gave more details on a point of contention between the two at a mid-September debate in Tampa: whether Perry had in fact increased taxes on Texans. Paul got in the successful laugh line that he'd critique Perry further, but feared that if he did Perry might "raise my taxes." Paul still runs two businesses out of

Texas, and says his corporate taxes in the state have definitely risen, and a lot, under Perry.

We chatted about mutual acquaintances from the libertarian world, including "crisis investing" guru Doug Casey, an anarcho-libertarian who made his name with best-selling books in the early 1980s days of feared economic apocalypse from government currency mismanagement. One of the critiques of this strain of libertarianism is that they've been foreseeing imminent danger and collapse forever—"predicting ten out of the last five crises," as one variant of the jibe goes—and thus deserve no credit for prescience if the fiat money game truly is over now.

Casey is a great libertarian bon vivant, using his wealth and the access it gives him to live out libertarian ego dreams such as negotiating with third-world tinpots to carve out free zones, insulting Dick Cheney harshly to his face, and hiding himself away in Argentinian jungle redoubts. Someone shares a tale of a mutual friend's troubles reaching a Casey hideaway in some South American mountain land. There must be easier ways to find freedom in an unfree world, Paul jokes, referencing the title of the famous 1970s bestseller by Casey's fellow libertarian financial doomsayer, and later LP presidential candidate, Harry Browne.

We talked about Vince Vaughn—he's getting more knowledgeable on the issues and harder core about war and foreign policy issues as he gets more involved, Paul said, and he might make a good candidate himself someday. In preparation for the convention panel on medical regulation, Deist and Paul explained to me the specifics of a bill Paul had sponsored that would eliminate an antitrust restriction that prevents doctors from joining forces to negotiate with insurance companies over prices.

Paul shared some gossip he's heard from people involved in border control about the realities of the "border fence," the upshot of which is that it isn't doing any good and is probably not even designed to. Anyway, as Paul explained at the Reagan debates, even the most bloodily efficient and tyrannical of border fences might work, but it wouldn't be American. Besides, do we really want our country surrounded by an impregnable barrier that could work as efficiently at keeping us in as it does keeping them out? The level of controls over the movement of property, cash, and capital out of this country are already extreme. Jon Stewart riffed off this one in an affectionate bit positing Paul as our national Kramer—the zany neighbor with a skewed view on life and an endless parade of unworkable and weird ideas, some of which do inexplicably work. "Ron Paul is the only one who has realized the day may come when we might want to sneak into Mexico!" Paul thought the line was hilarious.

We landed at Burbank just in time for Paul to make it to his slot speaking at the GOP convention's Lincoln Club breakfast, two hundred people crammed around tables in a room barely big enough for them.

He was roundly cheered, and mobbed as he tried to work his way out the front. A member of Paul's security detail told me they angered the hotel security when their car arrived at the wrong door, and were henceforth left to their own devices; lacking knowledge of the back ways, Paul was just led back and forth through the hotel's corridors, followed by a mob of four hundred or more young fans, many of them bellowing "End the Fed!" and "Ron Paul" or "President Paul." This was, of course, the only sign of life the California GOP convention showed.

"President Paul" was a deliberate bit of word/attitude magick via yoga teacher and Paul activist Steven Vincent. If we want to believe Ron Paul can be president, we have to get used to saying that Ron Paul *is* president. It's like verbal visualization. Vincent thinks hearing it will help steel Ron himself for the task ahead of him. He thinks Paul and his fans have to begin sincerely believing and pronouncing that he can and will win, that this is more than just a game of ideological promotion or a quixotic attempt to toss unusual ideas into the public debate. And he thinks hearing the words "President Paul"—from their lips to Paul's ears—will help.

Floating along in this mob, especially for someone who can remember a quiet Paul with a quiet crowd of around eighty on a college campus twenty-three years ago—was quite literally dreamlike, or even seemingly fake: surely this couldn't be really happening. Many hundreds, mostly under thirty, white, Hispanic, Asian, mohawked, bowler hatted, suited, T-shirted, with their homemade signs and the campaign's printed signs and their own printed signs and a galloping joy and enthusiasm, rushing, marching, zipping forward to get a glimpse or photo of their man from the front, falling back with a smile, in fellowship and passion—it seemed impossible.

But as cranky Republicans pointedly slammed the doors of their nearly empty meeting rooms, as non-Paul delegates just stepped out of the way and gaped or muttered sour little insults to themselves, it was happening. It was real. It was so real it forced Paul to just give up going into the small room where the medical panel he was expected at was being held; it was already two-thirds full but fit no more than one hundred at best, and he couldn't drag this mob in. On the fly, the decision was made that the peace and progress of the panel would be better served by Dr. Paul not participating. He just got shuffled through some "authorized personnel only" doors and disappeared from view of his people until his scheduled speech before the Republican Liberty Caucus an hour later.

I sneaked into the room where his speech was to start in fifteen minutes with my media pass while hundreds of Paulistas were crowding outside. Paul was shooting a TV interview up front. Benton was visibly drained by the stress of the roaring passionate crowd, and wondered if that sort of thing on occasion does more harm than good to the *real* cause right now: not just the pleasure of expressing one's passion for Paul, but convincing the not-already-convinced that this is the political wagon they wish to hitch themselves to.

Paul was slid out the back door to await his stage time, and the front doors were opened. Shawn Steel from the Republican National Committee gave a quick introduction, intended to assure this shouting rabble that they had an honored place in the Republican Party as he sees it. You have brought, he told them, a raw energy this party had not seen in decades, and lowered the average delegate age by thirty years. He praised Ron as a "dignified, elegant, wonderful human being who gave us his son and improved the Senate's DNA," and assured us that as long as he's around, the Paul folk and "lovers of freedom" will always be welcome, always have a major role in the Republican Party.

As Paul hit the stage, the crowd went wild. He brought out that Paul-in-concert rarity, the call for an end to the war on drugs. Paul's firmness on this stance is unquestionable. But as he drifts and riffs through his grab bag of ideas and expressions in his talks, it doesn't always come up.

The crowd seemed awash with their own triumph, the sheer improbable spectacle they'd summoned: it's like Ron Paul himself was some apparition created by their own mental energy, their own time sign making, talking, posting on the Internet. Steven Vincent took the lectern earlier, and he stood with a tight smile on his face and his head bobbing as if he were feeling their enthusiasm and energy working up from his own toes and he was carefully keeping it in check, channeling it for later use.

Then it was down to business. They needed to project their love for Paul beyond just the stunned and annoyed California Republicans in the Marriott. Later that day, a day in which no one could have been unaware of how loud Ron's fans are, of how they constituted at least a third of the total attendance at this state party convention, I sat through a very well-attended panel of political experts talking about the 2012 presidential race. More than one hundred California Republicans were in the room. The name Ron Paul was never mentioned, not even to be snidely dismissed. They'd heard his name loudly and often enough today. It was not going to be passing their lips. Paul's fans didn't even bother infiltrating this panel. But this convention really felt more like some boring Republicans cluttering a Ron Paul rally than it did a Republican convention. It was a Ron Paul convention. Paul won the straw poll, 44 percent to second-place Rick Perry at 29.

Having walked up and down the straw poll voting line four times during the day, each time with around thirty people in line, I expected the Paul victory to be even more smashing than it was; I don't think I saw more than ten people *not* festooned with Paul flair or holding Paul signs ever in the voting line. But the vote count doesn't lie, I guess. (Watching out for possible shenanigans—taking away boxes of votes, pulling in new boxes that might already have votes in them—was a high priority for Paul's volunteer grassroots poll managers.) I chatted in line for a while with an Iraq vet there with his wife and infant daughter. He understands the importance of Paul's foreign policy very well and dragged his family up from Orange County to make sure his support is counted.

Paul's victory in the California straw poll was barely noted by the political media; it certainly didn't shift the narrative of the campaign. A week later, in a process far more easily gamed by a set of party insiders, Herman Cain won a Florida GOP convention straw poll with 37 percent, over Perry's 15. (Paul got 10.) This instantly propelled Cain to front-runner status in media accounts, which actually began being reflected in polls.

Ron Paul was still running, though. This time, on to New Hampshire.

New Hampshire vigorously defends its traditional status as the first primary election in the nation (Iowa's is a caucus). While I made an extended visit in late September and early

October to cover a Paul campaign swing through the state, Florida announced, at the cost of the loss of half of its delegates, that it was moving its primary election to January 31. New Hampshire just pulled up its belt and decided to move its forward as well, settling eventually on January 10.

This added an almost impossible element of haste to one of the Paul grassroots' efforts to pump up Paul's poll results, the "Blue Republican" campaign to encourage Democrats, given their own lack of any interesting primary action this go-round, to (as they can do in New Hampshire) change registration to Republican and vote in that primary for the antiwar candidate, Ron Paul. But they have to make the switch ninety days before the primary and that's suddenly sooner than anyone anticipated.

Because a victory, upset, or surprisingly good showing in New Hampshire can mean a lot for buzz and momentum and money, candidates fight hard over the state. Thus it's one of the only places where politicians can be relied on to show up in people's backyards if invited. Our hosts tonight were a pair of Paul supporters—married couple Michael and Erica Layon. He's in medicine, she's in finance; they raise their own chickens and sky-dive for fun. They promised a "leap for liberty" to accompany this backyard event—it would be a daring husband-wife helicopter jump, from which they'd stretch between them a seventy-five-foot-long Ron Paul banner.

They did say "weather permitting," and it did not permit; it was a damp chilly night. By the time Paul took the podium, the hundred or so of us huddled under white tents in their backyard were sliding in, away from the edges where the water pounding the tents dripped down. A green laser painted the words "Ron Paul Revolution" on the inner roof of the tent.

Rick Perry announced his candidacy at a similar New Hampshire backyard party, Buzz Webb told me. Buzz is a lesbian Ron Paul activist; she used to want to live only in "gay meccas" until she discovered the freedom movement through Ron Paul. She now can't imagine being happier anywhere other than surrounded by other anarcho-Paulians in the Free State of New Hampshire. A gang of Paul activists found out where Perry was announcing. Of course, they grabbed their Ron Paul signs and surrounded the house.

Paul arrived unceremoniously in a minivan, his usual security absent. Before the talk, Ron worked the crowd—a crowd with a handful of state legislators there to demonstrate their respect. A younger man, also a doctor, began connecting with Paul over their mutual profession. Oh, you picked a great life, Ron told him, with beaming enthusiasm. Medicine used to be wonderful, Paul reminisced, before the government took it over. It used to be that doctors would charge the least they could; in a bureaucratized insurance-government duopoly strangling the entire industry, it's all about charging the most anyone can charge—everyone is a step away from the person actually paying. The young doctor mentioned his favorite memory of payment—a sack of corn and a watermelon.

Ron's introduction was a triumphant moment for the Free State Project and the larger cause of the freedom movement in New Hampshire. We, the freedom-loving New Hampshirites, helped elect one of them as Speaker of the New Hampshire House, Bill O'Brien. The legislature under his leadership achieved something the state has failed to do for years: it actually cut the budget, by 11 percent.

O'Brien introduced Paul in a way that would be news to most of the party's elite: as the very definition of Republican values, who understands that it is family that instills values. (O'Brien ended up pissing off Ron Paul fans in New Hampshire by endorsing Newt Gingrich.) As the tent bulged threateningly over us, Paul started with a joke about the aborted jump—"they told me I'd have to skydive in, but . . ."

He admits he can sound a little depressing, stressing the economic mess we are in. But when he transmits this message of looming econopocalypse to the young, he says, they come out strangely buoyed: because he is helping them understand why what is happening is happening, and to grasp that solutions exist, that an end to debt accumulation and a return to a solid currency will allow the natural energies of unrestricted markets to start creating wealth again, as they always do.

He doesn't always hit the drug legalization note, but here, where he was supposed to be selling himself to theoretically undecided Granite State GOP primary voters, he did. (I can detect no pattern of strategic logic at work dictating when Paul pulls in some of the rare cuts from his speech repertoire. Money, spending, the wars, he pretty much plays every night, the inescapable hits of his repertoire. The areas where his thoughts veer farthest from the libertarian and closest to the right-wing populism from which he arose, such as abortion and immigration, he almost never brings up in his own unprepared remarks, though he will engage them when asked.)

It was an extension of his controversial riff on heroin legalization during the first GOP candidate debate of the season, in South Carolina in May. We tolerate people reading whatever books they want; we are fairly tolerant of whatever religion people choose. But when it comes to *bodies*, well, we tend to think the government should be able to crack down and enforce its rules, as if we don't have the brains to manage our own affairs. He's being subtle here, taking people by the hand and walking them from someplace he knows they are comfortable to someplace he figures they probably aren't, gently helping them see they can stay on the same trusted path on the journey.

During the question period, in which Paul was asked to mediate between a daughter and father on whether libertarians have to be antiunion (no, as long as the unions aren't coercing anyone)

and a Hispanic man wanted to know how to explain Paul's marijuana stance to his mother, a mother handed the mic to her son,
who looked to be around four years old. He said "thank you" to
Paul. Paul cheerily said back, "You're welcome."

There were more questions: on immigration, the use of special
forces, some special pleading for disabled child health services.
(Paul never seems willing to just tell people, yes, I know someone is benefiting from every government giveaway; just because
someone is benefiting from it doesn't make it right, or sustainable
public policy.) Then he began the slow walk through the crowd.
He got pulled back onstage to pose with a group of skydivers,
holding a copy of *Parachutist* magazine. Paul, unbidden, engages
them in a minute of serious questions about the mechanics of helicopter jumps versus airplane jumps. This is the guy his fans wish
everyone could see, not just the harried, often angry-seeming hec-
torer who arises when his core beliefs are attacked, surrounded by
crooks and enemies, at GOP debates.

A very dedicated volunteer named Leah Wolczko grabbed
Paul and told him he has to stress more often that he's the guy
who saw the economic crisis coming, who understood why it
would come. Well, Paul explained, it isn't really in him to be like,
oh (in a pompous, braggy voice) "Oh, I warned you!" Sure, his
campaign staff tells him he should, but it just wouldn't be him.
But you've got to, Leah said, you've got to. Paul saw that he wasn't
going to win this one—"You're right!" he told her, with a smile.
Then you'll do it? She was thrilled. "I didn't say I'd do it—I just
said you're right!"

It was a typical Paul walk; told a guy to read nineteenth-
century French libertarian writer Frédéric Bastiat's *The Law*;
asked an undecided guy what issue he's hung up on; met a guy
who came in from the Netherlands to volunteer for a campaign
for president for a nation not his own. Paul looked at me. He had
taken to doing a comic double take every time he saw me lurking

around, like, "You again?" He noticed I'd gotten a haircut since he last saw me in Los Angeles. He told me Rand used to cut his own.

The next morning Paul appeared at a breakfast event at St. Anselm's College, a rambling campus run by Benedictines. This was not a Paul audience—it was full of representatives of the New Hampshire establishment, government, law, insurance, including Comcast, Fidelity Investments, Savings Bank Life Insurance, Harvard Pilgrim Health Care, and Liberty Mutual. As usual, there was no sign that he had crafted the speech for the particular circumstances of his audience. Before he talked, he worked the room, shaking hands with nearly every one of the hundred or so in attendance. The caterer saw him, and told me that she thinks Paul looks "so fragile" as he moves his wiry way around the room. Not at all, I told her; he's still capable of twenty-mile bike rides regularly. That very afternoon he went on such a ride with a reporter.

In the question-and-answer session, he told this veteran-filled crowd of New Hampshire grandees that to believe that Muslims want to kill us merely because as Muslims they hate or fear our freedoms is just not correct. He pointed out that through his career in medicine he's known many Muslim doctors—fine doctors—and he's quite sure that they too are perfectly happy to be wealthy and free. (The doctor who helped his wife during her 2007 campaign health crises was a Muslim.)

He got a local question, a tax question: Why does Massachusetts apply its income tax both to those who live there and those who work there and live elsewhere? What would President Paul do about this? Well, that's a state thing, he said. He was not going to offer to solve this guy's problem. He may not approve of that tax policy of Massachusetts, but it's not a matter for the federal government.

Outside there was a small press scrum. It's the morning that

Barack Obama ordered the summary execution of alleged al-Qaeda collaborator Anwar al-Awlaki. Paul was asked for a reaction. Ever the constitutionalist peacenik, he was against it. "Nobody knows if he killed anybody," Paul said. If the American people just accept this assertion of imperial power, that's just sad. As usual, he was the only Republican candidate who thought this. Except Herman Cain, the first time he was asked, but he was the "other front-runner" of the moment (Anyone But Ron Paul!) so he quickly backpedaled. A country that had been riven four years earlier between those who believed George W. Bush had the plenary authority to detain and waterboard suspects in the war on terror and those who thought that was the mark of a tyrant had been united: the Democrats were now happy with the tyrannical powers in the hands of one of their own, and Republicans continued their love affair with presidential death dealing under whatever brand.

Paul was asked his opinion of the Occupy Wall Street (OWS) movement, then in its first media flush. The OWS crowd has complaints similar to Ron Paulians' but with a very different set of proposed solutions. For the protest's triumphalists, this was the True Revolution, the beginning of the end, the collapse that Paulians suspected was coming. Paul had been including hat tips in his speeches for weeks to the notion that the chaos of the Greek bond collapse could presage actual unrest and violence in the American street. Was this it?

Paul talks very slowly and carefully in these press scrums, as if he feels someone is trying to trip him up. He admitted that he isn't intimately familiar with all the Occupy groups' demands, but "if they were demonstrating peacefully, and making a point, and arguing our case, and drawing attention to the Fed—I would say, good!"

The supporters of the man who wrote the 2008 bestseller *The Revolution* were not shaping this assault on Wall Street. The Occupy Wall Street moment defined the dilemma of the Ron Paul movement vis-à-vis the stronger currents they moved in: four years of government failure and malfeasance later, more people were willing to blame capitalism and markets and beg for more government solutions than to seek the disengagement of government from business and the benefits of a fully voluntary society.

Paul told me in October that he does recognize in the Occupy movement "a tremendous opportunity" but that running in a GOP primary, "it is not necessarily advantageous to overemphasize alliance with people the conservative voters don't really want to talk about. They know I associate with Barney Frank on cutting military spending, and on marijuana; even in my own district, that hasn't hurt me at all. But to go to Wall Street now . . . I don't endorse everything everyone is doing up there. It's a mixed bag. Some don't like capitalism and that's not my beef. I've been the one very particular in defining free markets versus the crony capitalism."

The feeling I got from Paulians in New Hampshire and Los Angeles who visited Occupy groups was a strong sense of sympathy with the energy, the will, and the enemies of these protesters, combined with frustration that they just didn't understand economics well enough to realize that it was not capitalism that was their enemy, but rather crony capitalism—"crapitalism" in a neologism floating around at the time. Steven Vincent in Los Angeles spent a lot of time down at the Occupy demonstration, and even got involved in their tortuous consensus procedure in order to get ending the Fed added to their list of demands. He said there was a lot of agreement on that point. But in Philadelphia, hostility on the part of the Occupiers to the Paulites who joined them got so extreme someone broke into the Paul tent, stole all their literature, propane, and most of their food and water, and left human excrement on the floor.

It was just like Paul's old hero Leonard Read thought:

economic education, mind to mind, was what would change the world. But another of Paul's great influences knew something Read didn't. Ayn Rand, the imperious Russian novelist, thought almost no one her equal, or even equal to her task: changing the world's attitudes about reason and capitalism and liberty. But for a while she thought Leonard Read was different. She became disillusioned when Read dared publish a pamphlet about rent control that Rand found dangerously red—a pamphlet cowritten by that notorious enemy of free markets, Milton Friedman.

What enraged Rand about Friedman's pamphlet was that he was willing to grant the good intentions of his intellectual adversaries. Paulites were generally doing the same with the Occupy protesters. After all, they agreed Wall Street deserved condemnation. To the Paulite, the Occupy Wall Streeters blamed on property what was really the fault of politics, and wanted from politics what could only justly be gotten from property, justly shaped by the free market. And the more the Occupiers noticed Ron Paulers among them, the more it seemed to annoy them.

The people of the Free State Project were still, most of them, doing activism for Ron Paul. Free Stater and state representative Jenn Coffey, when asked about efforts she was making to campaign for Ron Paul, contemplated her life and friends and job in the state house and her eagerness to talk and promote the cause of liberty anywhere and everywhere and answered: "When are you *not* campaigning for Ron Paul?"

The New Hampshire Ron Paul crowd, across the state, was unusually close-knit, and more inclined toward anarchism and civil disobedience than most others I encountered, largely thanks to the energy of the Free State Project, which has worked for nearly a decade encouraging the libertarian-minded to relocate to New Hampshire. The president of the FSP, Carla Gericke, believes they will succeed in changing the political and regulatory climate in the state to turn it into the "Western Hong Kong—make it a really business attractive environment, so people will think, Let's start gold banks, build everything ideologically we think should exist in the world, build it here." Working for Ron Paul fits in to their larger vision of shaping the ideology of the state and the nation in a libertarian direction, even for those among them who have nothing but disgust for government. Gericke mentions that their comrade who goes by the name of Sovereign Curtis affectionately calls Ron Paul his "favorite government thug."

Loosely associated groups such as CopBlock (dedicated to exposing and fighting police misconduct and brutality) moved in and among the Free Staters. Personal stories of arrests for civil disobedience from chalking sidewalks with antipolice slogans to filming cops or judges abound among this crowd. Simultaneously, a dozen Free Staters inhabit the state House of Representatives. State Representative Mark Worden told me, in a group interview with some other Free State/Ron Paul activists filled with tales of warning motorists of looming DUI checkpoints (police hate that) to how their community hangs together supporting each others' myriad small businesses, that as near as he can tell he knows more anarchists than he knows Democrats. A lot of the Free Staters don't believe in electoral politics at all but still believe in promoting Ron Paul, often through old-school efforts such as hanging huge guerrilla signage on freeway overpasses and the like. But as William Kostric tells me (Kostric was a brief media star in 2009 for showing up at an Obama rally with a legally carried gun), he'd be perfectly happy leaving all this political activism hubbub behind if the Ron Paul thing doesn't work, that he could be happy just living on a beach in some warm foreign clime and fishing. "The country is going down the toilet, and this seems like one last chance," he says. "But I do not want to let it be said that I did nothing."

Still, the New Hampshire operation, being run again by Jared Chicoine in 2011, had plenty of the more traditional phone banking he treasured. Even back in October, he told me they had enough volunteers to make over 2,000 calls in a night; "People on their way to a doctor's appointment will stop in [at the campaign's Concord office] for a half hour and make calls."

Ron Paul has a unique hold on his fans. We all see him as standing for us, as the sole representative of an eccentric and derided political outlook who has any hold on public and media attention. And he's not always speaking as well as we would, or at least not exactly as we would.

Watching Paul give speeches or debates along with other Paul fans, we are often cringing, or at least rewriting his talk on the fly. Paul resolutely refuses to go through anything like rigorous professional debate training or preparation. He has been talking about this liberty stuff for decades the way he likes to, and wants to keep doing it. But Ron Paul *is* us, is Ron Paul's fans, in the public eye, and we desperately want him to explain and defend our beliefs with perfect cogency, be perfectly convincing. When he embarrasses us or stumbles, it cuts as deep and personal as when a child or spouse disappoints us and embarrasses us. He's ours; he's us; he's our representative in ways he could never be to his actual Texas constituents. And he is resolutely uncoachable; he refuses anything like disciplined debate preparation or rehearsals. As Jeff Frazee of Young Americans for Liberty, who traveled with Paul in 2008, says, "Ron's attitude is, 'I've been talking about these ideas for thirty-five years and this is who I am, this is how I present my message.'"

And he's objectively very good at it—just count all the devotees. Still, I can feel distinctly uncomfortable sometimes rewriting Paul's own talks in my head, and I definitely felt that way as I sat in the cafeteria of Lincoln Financial Group—Concord's third-largest employer—and watched Paul get hit with the most difficult audience question I'd ever seen him get. It came from an almost aggressively earnest man with a long face and a gray beard, exuding "stressed-out old liberal," the kind who is just driven to distraction with how he just can't understand how some people can be so ignorant. Mr. Paul, do you consider these government functions justified? He had a list. A long list. He only got three of them out, but he had ten.

This was it. The central conflict of the libertarian versus, well, everyone else. The unique difficulty Ron Paul can face on the hustings, both from those just clever enough to skewer Paul over his quirky political philosophy, and from those who just sense that they don't understand where this guy is coming from.

Everyone understands where Barack Obama is coming from. Everyone understands where Mitt Romney is coming from. They are just coming from, well, where we are. They want to keep doing what has been done. They express it in slightly different terms and will insist with varying degrees of venom that the other side isn't doing it right, or fast enough, or smart enough, or frugally enough, or in a way that punishes the right people. But everyone gets it.

But to understand Ron Paul, you must understand a complicated set of both fact and value propositions, about the real benefits of government programs, and what might be done with those resources if government didn't commandeer them, and the history of benevolent societies, and the percentage of tax monies lost to bureaucratic inefficiencies, and the possibilities of user-feeing certain things, and balancing the costs of good medicines withheld versus bad medicines available, and how free-market regulatory authorities like Underwriters Laboratories work, and why the long-term value of everything is best calculated via allowing owners to bear costs and reap benefits, and . . . a whole Libertarianism, not 101, more like 303. And this guy wanted it in sixty seconds. He wanted to have the entire intellectual edifice of his life torn down and a sturdy new one built for him in a minute, on the spot. And he didn't get it. And I wished I could leap up and take over for Paul, for those agonizing seconds. If you know what he's talking about, you can always tell what he's talking about. But I'm not sure random non-libertarians do. It was a long day for Ron. The campaign sent him where they were invited, where they thought it would do good. Rallies for the fans are fun and all, but that was not what the campaign needed, months before the primary. They needed to get influential people who didn't know and love Ron Paul to know and love him. Thus he was appearing during activity night at the Haverford-Heritage Retirement Home before an audience of fifty or so New Hampshire elderly—the elderly, whom you can generally rely on to vote, and to be thought leaders, or try to be, among their friends.

As at Lincoln Financial, it didn't seem that many of them came in as fans. One man, Ralph Bennett, sitting at my table, was very familiar with Representative Paul, though. In fact, he told me that for ten years he had been writing to Paul regularly, advising him to get the United States out of the UN, and the UN out of the United States. I asked him if he knew that Ron actually does introduce a bill to get us out of the UN pretty regularly? Ralph shot back the number of the bill to me instantly.

Paul beamed indulgently at the hesitant-voiced ninety-two-year-old woman who gave him a worshipful introduction, stressing the wonderment of all his achievements as a military man, doctor, and statesman. He did one thing at the start that seemed like situational pandering—stressing the importance of family, something that I'm confident is true for Paul, but also something he seems a bit too restrained and private in an old-fashioned way to toss out carelessly for political gain. But he let this older crowd know about his family, and told a little joke about Carol's leap-year birthday, and his supposed attempt to get away with buying her a birthday gift only every four years.

He was certainly not pussyfooting around when it came to the questions, though, and it was a reasonably tough bunch. One gruff veteran told Paul that he loved the veterans ad. It was Paul's first heavily placed campaign commercial in New Hampshire, one that elided any of Paul's unique policy positions but used actual vets from his district to talk about how good he is at honoring them and getting them medals they deserve. But, the man grumbled, Paul needs to stop running it so damn often, he feels like he's seeing it every fifteen minutes.

Paul was challenged on his foreign policy and did not retreat, and in fact riffed off into his Iranian history lesson and explained why he thinks it's a silly lie to believe that they hate us because of our freedoms. In one of those locutions that you'd hear from no other American politician, the kind that makes his admirers feel their admiration turning to sweet mushy love, he said he wants to "reject the notions of violence and destruction" when it comes to politics.

That veterans ad was vital to the campaign's early messaging, because Ron Paul's biggest problem is his foreign policy—even though in another sense Ron Paul's greatest attraction is his foreign policy.

Both statements are very likely true. Pretty universally, when I ask someone in the amateur or professional business of selling Ron Paul to potential voters, What's the biggest stumbling block they run into? the answer is: his foreign policy.

That was made abundantly clear at that moment in May 2007, the dustup with Giuliani, that launched the revolution in earnest. Paul's foreign policy puts him outside the normal realms of not just his party, but any party. He's the only politician willing to judge America's foreign policy adventures by the same moral standard we apply to other countries' foreign policy adventures. As he said frequently on the 2011 campaign trail, we should adopt a Golden Rule of foreign policy: doing to

other countries only what we would want other countries to do
to us. He'd dare apply empathy to our judgments of how other
countries behave when we go over there and bomb the hell out
of them. As Paul describes the gamut of American treatment of
the rest of the world, we either treat them as our puppets or we
bomb them. He understands that history doesn't begin when we
decide it begins, that there are facts of past U.S. behavior that
help explain the foreign policy conundrums we are in today.
That's what the blowback brouhaha with Giuliani that started
it all was all about.

He's aware, he has to be, that his foreign policy stance is his
hardest sell with the modal GOP primary voter. So he frequently
reminds his audiences that a humble foreign policy is a political
winner—for candidates. Alas, it's a stance that politicians tend to
abandon once they win the voter's favor.

Forget the now-ancient history of misleading chief executives
such as Woodrow Wilson and Franklin Roosevelt, both elected
on promises to keep us out of the European wars that they then
rushed us into. Eisenhower and Nixon were both elected on
promises to wind our way out of unpopular Asian wars that their
Democratic predecessors had bogged us down in. Even the last
GOP president, George W. Bush, prior to starting two new wars
followed by endless nation-building projects, had run and won
on the promise of a humble foreign policy. And for those whose
political memories were painfully short, he reminded us that our
last winning president, Barack Obama, despite keeping us so far
in all the wars he inherited and starting a new one in Libya, was
the peace candidate. His fans liked to turn the *O* in *Obama* into
a peace sign. The Nobel committee in Stockholm gave him a pre-
emptive Peace Prize for nothing other than the image he'd been
elected on. Peace sells, and when given the chance, the American people absolutely were buying. They just got continually
cheated by the political sellers. Paul was trying to let Republican

voters know that being for peace is a proven winner in national elections.

In October, Paul did a pretty un-Paul like thing, one he told me at the beginning of this campaign he didn't expect to do: he issued a detailed budget document. It wasn't detailed like an actual federal budget, not many hundreds of pages, but ten itemized pages listing $1 trillion in spending cuts immediately, and indicating a balanced budget in three years with no tax increases. It kills five departments entirely, moving any necessary constitutional functions elsewhere, but bye-bye departments of Commerce, Interior, Energy, Education, and Housing and Urban Development. He'd freeze the military at half a trillion and end all our current overseas engagements, and freeze the State Department at $7.9 billion.

The Tea Party–organizing FreedomWorks praised it highly as the only realistic fiscal conservatism arising from the GOP field, happy that he proposes "abolishing the Transportation Security Administration, ending corporate subsidies, eliminating the Death Tax and repealing ObamaCare, Dodd-Frank and Sarbanes-Oxley." The *Washington Post* found a bunch of economists to swear that cutting government spending would be economically disastrous, and even supposed conservative ones played the "we absolutely must cut spending, but not actually ever now" gambit. Paul also begins the process in his budget of allowing younger people to leave Social Security and Medicare, while ensuring that benefits for those on the programs remain at their promised levels. He'd encourage local experimentation, and slough off some federal responsibilities, by block-granting Medicaid, food stamps, and child nutrition programs. He cuts the federal workforce by 10 percent, but leaves his favorite government program, veterans affairs, funded at current levels. *BusinessWeek* attacked everything about the program except the lesson they think it teaches—that actually cutting federal spending is really just unthinkable to anyone, anyway.

Paul is lackadaisical about the document. "It wasn't the most important thing for me to do, but that's what people want when you talk about cutting spending: Are you serious?" He told me he could never get any of his debate opponents to specify any substantive cuts. "It's like they think you can cut child health care and a few entitlements and get rid of the debt, but don't you dare touch one nickel in the military! I'm trying to make the point that it is a serious economic problem we are in because of debt and you can't get out of the mess until you start really cutting."

Yes, Paul is serious. Paul is by all personal accounts exactly the kind of man the Christian right "values voter" should want in a presidential candidate: serious family man, devoted to one woman, successfully raised five children with many happy devoted grandchildren and even great-grandchildren in their wake, a serious Christian.

But he refuses to make a big deal out of the latter, even when he can, even perhaps when he should. I've seen him lecture to an audience of Iowa pastors, and while he is savvy enough to lay out some of his pro-life and border control ideas to them (which he doesn't when talking to college kids—Paul does not say what he does not believe, but he does choose what to say when, often with political care), he doesn't say anything about his own relationship with God or the church. Raised Lutheran, he married in the Episcopal Church and stayed in it through the raising of his children, but of late has attended Baptist churches, I'm told (though not by Paul himself), because of his discomfort with Episcopal looseness on abortion.

Paul knows his reticence isn't winning him evangelical fans in the Republican Party. "Some evangelicals get a little bit annoyed because I'm not always preaching and saying, 'I'm this, I'm this, and this,'" Paul said in an interview for the organization Faith in Public Life in 2008. "I think my obligation is to reflect my beliefs in my life. I like the statement in the Bible that when you're really in deep prayer you go to your closet. You don't do it out in the streets and brag about it and say, 'Look how holy I am.'"

He's not much of a public breast-beater about the sign that his foreign policy, as confusing and aggravating as it is to the GOP faithful, seems absolutely right on to a very relevant constituency: active-duty military, who give to Paul more than to all the other GOP candidates combined, well over $100,000 in 2011.

Before a town hall meeting in Nashua, New Hampshire, in October, Paul held a small press conference, with a line of sixteen veterans standing behind him, veterans of conflicts from Korea to Iraq. Paul explained his desire to only expend the lives of our soldiers in the direct defense of America. The men were members of Veterans for Ron Paul, one of dozens of special interest Paul fan groups deputized to bring his message to their specific communities.

One of the most famous Veterans for Ron Paul (not at that event) is Adam Kokesh. Before he was a veteran for Ron Paul, he was an Iraq Veteran Against the War. His libertarianism preceded his antiwar activism, and his Paul activism. "I got into libertarianism as a set of practical policies and as a unique self-identifier" during his younger days, "like, 'Oh, I don't have to be a liberal or conservative, a Democrat or a Republican; I can be a libertarian!' That was cool."

The Paul campaign in 2007 coincided with the height of Kokesh's antiwar activism with Iraq Veterans Against the War, a mostly liberal group. He decided to keep working in that world rather than becoming a full-time Paul volunteer; it seemed more important to push the Paul perspective outside its natural home, "repping Ron Paul to the antiwar movement."

As Kokesh's veteran cred and penchant for forceful public outreach—which frequently ended in arrests, though no convictions—made him a star in the Paul community, people began telling him he should run for office himself. By the end of 2008 he gave in to the entreaties, asked Paul himself for a rare endorsement for a congressional seat from New Mexico, and got it. (While Paul is still restrained in handing out official endorsements, and doling out cash from his LibertyPAC to candidates not named Paul, he can be light about it; when a questioner at a Nashua town hall meeting in October, whom Paul had never met, asked at the mic for Paul's endorsement for a planned state office race, Paul laughingly and graciously handed out his endorsement to anyone in that room who ever decided to run for office.)

Kokesh's run was a test of the nascent Paul movement; most of his staff were out-of-staters drawn from the ranks of the Paul revolution. He was able to rely on the tried-and-true moneybomb and Internet fundraising techniques honed in the Paul campaign for about two-thirds of his fundraising; he still was surprised at how much time had to be spent on the phone fundraising. And he wasn't really prepared for how much organizational and paperwork effort was involved in running for office, as compared to mere activism. That same activism was used against him in the race, in "guilt by association—like, hey, you worked with liberals [in the antiwar movement], you must be a liberal." That became a proxy for all the other things that a Republican primary voter might not have been comfortable with about a libertarian. Kokesh learned a lot—mostly that he spent a little too much on some fripperies and staff and events and not enough on campaign basics. "Just because you as a candidate are unique," Kokesh said, "doesn't mean your campaign should be organized in any different way or spend money on different things than any other campaign has to."

Kokesh can't say he'll never run for office again. He's still young, still a vet, and now he's more experienced in explaining and selling his politics after his year as a radio and later TV talk show host. He did a radio show out of New Mexico for a couple of years and then shifted to a short-lived talk show on the *Russia Today* network in 2010–2011, called, naturally, *Adam vs. the Man*, which he was shifting to an Internet-distribution model when we spoke in October.

Like many Paul activists, his attitudes about politics have shifted. He's a "voluntaryist" now and thinks that there are richer opportunities for cultural and political change toward a freer world than just running for office—or being an active Iraq Veteran Against the War. "It's still a good cause and I definitely support it," he says. "But I'd rather be waking people to the message of liberty than just helping individual soldiers have more rest time between deployments."

Kokesh seems to enjoy being a big name in the world of Ron Paul activism. "It's been crazy to see fans of mine from the TV show saying on Facebook that I could be Ron Paul's running mate. Shit, I'm not even old enough! And I don't want that responsibility. I don't want my finger on the button. I've got PTSD. That's probably a bad idea. But what I think would be cool is if I ended up being White House press secretary for Ron Paul. Deal with the media and explain the message of voluntaryism over and over; I'd be cool with that.

"But I've evolved in my thinking, deepened my understanding of what it means to be a voluntaryist, thanks to Ron Paul. And it's more than just being a libertarian. I don't think government as a phenomenon is going to be ended by an election. It will be ended by a paradigm shift. And if Ron Paul wins this election cycle, then the paradigm will shift. It wouldn't be good to elect Ron Paul if we ended up with fifty tyrannical state governments. And as long as people in American society in large proportions believe in using the force of government to impose their will, there will always be some form of oppression."

Like many anarchist Paul fans, Kokesh is sure the congressman is really with him. "Kids take that attitude, that political activism is inherently bad because it involves government. It's easy to say that. It's a little more difficult to take the nuanced position that Ron Paul has and I share—that you can believe in the ideal of a world without government, I call it anarcho-pragmatism as opposed to anarcho-capitalism. That's believing that you can be pragmatic about how we dismantle the idea of government, how we get rid of the use of force in society.

"So many people have become dependent on this system based on coercion and violence, and rely on government for their daily sustenance. If we simply eliminated that, it would cause chaos." And running for office as a voluntaryist, Kokesh thinks, is not a clear violation of voluntaryist principles; rather, "you are forcing people to take a position, inviting people to understand what it means to be a voluntaryist."

Paul's campaign continued to surprise through the fall and winter. For a few days in late December Paul was actually polling ahead of everyone else in Iowa, right before that state's January 3 caucus. Almost certainly as a result of sticking his head above the pack, controversy over the old newsletters returned to the *New York Times*, Fox News, and nearly everywhere Paul was discussed on the Web.

He continued to deal with it by repeating that he did not write them, and did not agree with the sentiments expressed. (Elements in his campaign were pushing for a stronger response—a big public speech on race, or even trying to name the specific people responsible for the writing. Paul continued to keep his own counsel on the matter.) Despite widespread assumptions on the part of both many fans and many enemies that this resurfacing scandal would scuttle the campaign, there was no objective sign the controversy hurt him in the short term. While he ended up coming in a very close third in Iowa with 21.4 percent rather than winning, this wasn't because his polled support dropped much from his frontrunner peak. He lost first place because his opponents pulled ahead, especially with the surprise last-minute rise of Rick Santorum.

Paul earned his Iowa results through months of diligent and effective ground game, with huge ad buys and huge influxes of eager young volunteers working the phone banks that Paul's campaign team sees as the surest path to victory. But he was also inching up toward the mid-teens in most national polls, was the only not-Romney candidate with enough proven fervent support that he could likely fight it out to the end, and it was clearer and clearer that Paul and Paulism represented the true, and growing, opposition to business as usual in the Republican Party, and in American politics in general. As campaign manager John Tate said, "What happens this election cycle is more about the long term. No single election will save or destroy the country in one fell swoop. I'm hopeful whatever the outcome our folks will continue to shape the Republican Party."

Indeed, the results in Iowa showed signs that Paulism, with its appeal to the young and to independents, might be key to the future of the GOP. Entrance polls for the Iowa caucus had Paul pulling 48 percent of those age 19–29, and 44 percent of independents. But this doesn't mean that the GOP establishment, or any political establishment, will feel comfortable with Paul or his fans in the near term.

Anyone with any skin in the game of the status quo—politician, pundit, or citizen—has to find it difficult to take Paul seriously. That so many citizens and activists *are* taking him seriously scares the establishment for good reason. Paul doesn't just represent an opposition politician, he represents a denial that "the system" makes any sense, has any justice, or is sustainable. This radical oppositionalism makes it easy to just write his fans off as nuts and a bit scary.

Newsweek started to get at this important aspect of the Paul phenomenon back in February 2010, noting that "tea-partiers, Paulites, etc.—seem less interested in finding practical solutions to Washington's endemic problems than in tearing down Washington itself. As the 2010 elections approach, this nihilistic feeling will only grow stronger."

That's because the radical solutions that the Paul worldview demands—an end to overseas military adventurism, ending government's ability to manipulate paper currency, severe cuts in spending on all the myriad income-shifting promises Washington has made the past 80 years—don't register as "practical solutions" to the establishment. They seem like nihilism, though they are actually a belief in the American Constitution.

Any standard Republican or movement conservative really *can't* take Paul seriously without massive cognitive dissonance. You mean, we really *really* have to obey the Constitution, we really *can't* keep borrowing and inflating forever? Signs of a significant number of politically active youngsters believing in Ron Paul, which first became obvious at CPAC in 2010 and have become more obvious ever since, are indeed a sign of an apocalypse for the world that most politicians and pundits know. If Ron Paul is right, then everything they know is wrong.

Even left-progressives disappointed with Obama's civil liberties and foreign policy began noting the good qualities of Ron Paul as 2012 dawned, especially compared with President Obama, on civil liberties and war. Glenn Greenwald of *Salon* made that case best: "it is indisputably true that Ron Paul is the only political figure with any sort of a national platform—certainly the only major presidential candidate in either party—who advocates policy views on issues that liberals and progressives have long flamboyantly claimed are both compelling and crucial. The converse is equally true: the candidate supported by liberals and progressives and for whom most will vote—Barack Obama—advocates views on these issues (indeed, has taken action on these issues) that liberals and progressives have long claimed to find repellent, even evil." That Obama chose to sign the National Defense Authorization Act that codified his power to detain any citizen for any reason with no trial at the end of 2011, with Paul the only national political figure condemning this move powerfully, gave credence to Paul's progressive appeal.

Paul's campaign and movement were in a strange and difficult position as the 2012 campaign moved on, one unusual in American politics. Although there was no particular reason to believe it would happen, no one could stop fretting in December and January about the threat of Ron Paul running third party and ensuring a second Obama term, even as he remained a strong second behind Romney in most polls leading up to the New Hampshire primary. While the people responsible for steering the Paul political machine, from his political director Jesse Benton to his son Senator Rand Paul, saw the future of the Paul movement as within the Republican Party, the non-Paul Republicans certainly sensed that Paul and his people were in the Republican Party, by default, but not necessarily of it.

Paul was the only Republican candidate who everyone knew was unlikely to endorse and support any other GOP presidential candidate that wasn't him. He was the one whose fans were understood to be dedicated mostly to Paul, perhaps not at all to the Republican Party. The key to their forward success would be proving to the Republicans that they were too large a minority to ignore—that Ron Paul in 2012 was Barry Goldwater in 1960, swamped by the Rockefeller/establishment wing of the party, but still the party's inevitable future.

The Paulites will represent that future for various reasons— the most important one being that the looming economic and debt crises seem to demand a radically Paul-like solution. But be-ing right doesn't necessarily get you anywhere in politics. Passion and commitment are also necessary. And there's a reason why Paul people have such passion and commitment: because what's at stake, in the world of Ron Paul's revolution, is so important.

I was reminded of that by the leader of the New Hampshire Veterans for Ron Paul, Joshua Holmes. He was the youngest guy there, a veteran of two stints in Iraq, the first with the 172nd Stryker Brigade combat team.

Holmes ended up in the military in a sense because of libertar-ian ideas. As a kid he marinated in the works of libertarian sci-ence fiction author Robert Heinlein, whose novel *Starship Troop-ers* convinced Holmes that our brightest youth owed it to their society, their country, to be in the infantry. He quickly learned Paul-like lessons about the costs of empire and what being an oc-cupying army was like; he loves soldiering still but can no longer be a pawn of the U.S. government.

Holmes is a feisty and typical Paul fan—he's now living a fun and option-filled life in the Free State surrounded by fellow Paulistas, anarchists, voluntaryists, congressmen, swordsmen, ac-tivists, a roiling and hilarious community that is making him very happy, and he can't imagine not being part of this world. He wants to run for the state legislature himself; he's a native of New Hampshire, a young vet, he thinks he'll be a shoo-in.

Totally without making a *thing* out of it—it's a story he was telling because I'm a reporter with a recorder and I asked, he had lots of other things to tell me that day about his buddies and girls and plans that he gave equal weight—he told me about the day he was rolling around in his unit's Stryker, a twenty-ton, eight-wheeled vehicle that would roll in after a tank had blown a hole in something, and it was supposed to be strong enough that it could carry in some infantry and they wouldn't get hurt, at least not until they got out.

"The reason they used them in Iraq is because they're effectively bomb proof," he told me. "The only time that I got significantly blown up in Iraq we got hit by a suicide car bomb that had somewhere between five hundred and twelve hundred pounds of explosives in it . . . and the car bomb blew up and landed on the other side of our vehicle. It completely cleared a twenty-ton armored vehicle. And the worst injuries we had were second-degree burns and some guys got a little bit of shrapnel in their face." One of his buddies looked "like he had been shot in the face with a shotgun loaded with birdshot. All of the skin on his face looked really messed up. I saw him once as he came down out of the hatch and I thought he was going to die. He was just covered in blood from my perception. And I saw him later in the day after he got out of the hospital and he looked like somebody had rubbed his face with sandpaper."

So, the car bomb blew the hatch open. Josh said that if he and three buddies with crowbars had had four hours to work on that hatch, they couldn't have pried it open. One guy had his hand in front of his face, so that saved his eyes. Another lost some hearing.

It's a story of his life and he tells it, just like he tells stories about arguing about politics in class at the community college he's in now, and about funny characters in his Free State community.

And I began thinking of the difference between the world he and I were in—sitting in his living room, talking—and the world where giant machines of protection and destruction are rolling around, built so they can, you know, roll in after the tank has blown through the wall of whatever, and this guy was there away from his family and mom (he didn't even want her to know he was in combat—you know moms) and his girl, and their relationship fell apart because of that, and someone was so mad that this fucking ridiculous horribly expensive machine full of hostile strangers was rolling around destroying roads in *his* town and *his* neighborhood and in front of *his* family that he troubled himself to figure out the ways and cobble together the means to find a car and hundreds of pounds of explosives and *blow it up* . . . I mean, what kind of world is that?

Seriously, *this happens*??

It totally happens. It happened to Josh and he's not at all emotional about it.

I mean, *who's gonna stop this?*

Judge Napolitano says that what's so important about Paul is that he helps "people recognize what government is doing, so that when they are angry, they know what they are angry about. The movement grows every year and sooner or later there will be some sort of economic collapse, and whoever picks up the pieces will hopefully be those who understand what caused the collapse. And sooner or later some Reaganesque character who can articulate the values of the libertarian movement will sweep enough support with him to acquire the power to return us to constitutional government. I don't know if it will happen in Ron's lifetime or my own, but it is a beginning at least. No, not a beginning—a continual progression."

One thing I think I learned talking to all the other candidates and staff and family and fans and followers and his son the senator is that no one, right now, seems ready to adopt and bear the burden of the fullness of Ron Paul, in all its fearlessness, all its humanity, all its commonsense consistency, all its weight of forty years of thinking and reading and talking about the full-spectrum meaning of liberty and real free markets anywhere to anyone just because you know that is the most important thing in the world to do.

At the impromptu press conference outside his breakfast talk at St. Anselm's in New Hampshire, the one where he gave very un-Republican answers about murdering al-Awlaki and gave guarded praise to Occupy Wall Street, Ron Paul was hit with some other random detritus of that week's politics, questions that wouldn't be meaningful at all even a week later, and barely were then. Fundraising figures, Chris Christie, that sort of thing.

Then a reporter poked at Paul with some ABC polling figures that showed him not doing very well. Politicians who aren't doing very well in the polls are obligated to do something, no? How will you change your message, Ron Paul, in order to revive your poll numbers? he was asked.

"I don't change my message," Paul said. It was not the snap of someone delivering a killer movie-hero catchphrase, a one-liner. It was just that slightly hesitant Ron Paul thoughtfulness. "I change minds."

ACKNOWLEDGMENTS

──────── ★ ────────

A journalist's first thanks deservedly go to his interview sub-jects. This list does not include everyone whose insights about Ron Paul helped inform the book. It will not, for example, necessarily include everyone I spoke to for just a few minutes at events or rallies and does not include those who were only willing to speak not for attribution. They also have my gratitude. Some of the interviews were conducted for earlier journalism about Paul and not knowingly for this book.

My thanks to: David Adams, Representative Justin Amash, Jon Arden, Kate Baker, Bruce Bartlett, Jesse Benton, Steve Bi-erfeldt, Kevin Bloom, Phil Blumel, David Boaz, Vijay Boyapati, Eric Boye, Bryan Butcher, Lonnie Brantley, Jonathan Bydlak, Louis Calitz, Vincent Campos, John Carle Jr., Tucker Carlson, Jared Chicoine, Dick Clark, Representative Jenn Coffey, Neal Conner, Eric Cooper, Stanton Cruse, Christopher David, Jeff Deist, Bill Domenico, Eric Dondero, William Eddy, Mark Elam, Kevin and Marie Fiscus, Senator Jim Forsythe, Gary Franchi, Jeff Frazee, Eric Garris, Carla Gericke, Jackie Gloor, Charles Goyette, Anthony Gregory, Jeff Greenspan, Kelly Halldorsen, Ernest Han-cock, Nick Hankoff, Trent Hill, Joshua Holmes, Mike Holmes, Allen Huffman, Jack Hunter, Drew Ivers, J. Bradley Jansen, Dmitri Kesari, Dave Keagle, Christa Keagle, Matt Kibbe, Adam

Kokesh, Ann Koopman, Roger Koopman, William Kostric, Justine Lam, Chris Lawless, Mitch LeClair, Senator Mike Lee, Michael Maresco, Representative Glen Massie, Daniel McCarthy, Sandra McLaughlin, Brynn Menkhaus, Tara Menkhaus, Jan Mickelson, Zoe Miller, Judge Andrew P. Napolitano, Grover Norquist, Trygve Olson, Jim O'Neill, Senator Rand Paul, Representative Ron Paul, Roger Pruyne, Lew Rockwell, Chris Rye, Peter Schiff, Norman Singleton, Brinck Slattery, A. J. Spiker, Britton Sprouse, Bill Steigerwald, John Tate, Bill Toner, Robert Vaughn, Richard Viguerie, Steven Vincent, Robert Vroman, Kerry Welsh, Jon Watts, J. Buzz Webb, Nick Weltha, Richard Williams, Jason Wohlfahrt, Leah Wolczko, Tom Woods, Representative Mark Worden, and Ed Wright.

It can't excite anyone to hear someone is writing a book about them, and I appreciate the graciousness of Representative Ron Paul himself and various people associated with his congressional office, presidential campaign, the Campaign for Liberty, and Young Americans for Liberty who facilitated my presence at events, appearances, or interviews, or guided me to useful information or sources on the world of Ron Paul. I hope I am not forgetting anyone when I thank Rachel Mills, Gary Howard, Megan Stiles, Brian Early, Kate Schackai, and Teri Capshaw.

Thanks to Wlady Pleszczynski at *American Spectator*, who gave me my first opportunity to write professionally about Ron Paul back in 1999, and to the Competitive Enterprise Institute, whose Warren Brookes Fellow I was when I did that first writing. Thanks to my colleagues at *Reason* magazine, where I've done most of my reporting on Ron Paul, especially Matt Welch, Nick Gillespie, Jesse Walker, and Tim Cavanaugh.

Lucy Steigerwald, then an intern and now a colleague at *Reason*, provided ace research assistance in a hurry. *Reason* interns Seth McKelvey and Tate Watkins provided quick and efficient interview transcription services, as did Jessica Reeder. Thanks to all the posters and commenters at Ron Paul Forums (ronpaulforums .com) and the *Daily Paul* (dailypaul.com) whose information and gossip and arguments deepened my understanding of the Ron Paul story and phenomenon as it unfolded.

Various friends who might never care to hear the name Ron Paul again gave personal succor during this hermitlike process: my gratitude to Emily Brecht, Sarah Sanguin Carter, Tayler Jones, Joe Matt, Kestrin Pantera, Meghan Ralston, Al Ridenour, Heather Schlegel, Daniel Browning Smith, and Charlie Smith and the entire crew and family of the Fire Birds of the Fifth Direction in Black Rock City, Nevada. Thanks always to John Rinaldi for everything, and in the case of this book particularly, for standing up for the idea of Ron Paul. My wife, Angela Keaton, was always encouraging and supportive, and my apologies to my parents Frank and Helene Doherty and my brother Jim and his family for neglecting them while finishing this book.

Thanks to the Coffee T Bar on Sunset Boulevard (R.I.P.) and the Starbucks on Melrose and Stanley in Los Angeles for letting me sit around all day and night writing this book.

My agent, William Clark, is the greatest, and ferried this project from vague idea to contract to finished product with great skill, speed, and his usual sustaining calm and encouragement. Thanks to my editor Adam Bellow of Broadside Books for seeing the merit in this project and sheparding it to completion, and thanks to Kathryn Whitenight for her editorial work. As always, anything wrong with the final product remains my responsibility.

★